Problems and Solutions in Special Relativity and Electromagnetism

Problems and Solutions in Special Relativity and Electromagnetism

Sergei Kruchinin

Bogolyubov Institute for Theoretical Physics, Kiev, Ukraine

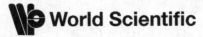

World Scientific

EW JERSEY · LONDON · SINGAPORE · BEIJING · SHANGHAI · HONG KONG · TAIPEI · CHENNAI · TOKYO

Published by

World Scientific Publishing Co. Pte. Ltd.

5 Toh Tuck Link, Singapore 596224

USA office: 27 Warren Street, Suite 401-402, Hackensack, NJ 07601

UK office: 57 Shelton Street, Covent Garden, London WC2H 9HE

British Library Cataloguing-in-Publication Data

A catalogue record for this book is available from the British Library.

PROBLEMS AND SOLUTIONS IN SPECIAL RELATIVITY AND ELECTROMAGNETISM

ISBN 978-981-3227-26-2
ISBN 978-981-3227-27-9 (pbk)

Typeset by Stallion Press

Email: enquiries@stallionpress.com

Printed in Singapore

Foreword

Field theory is an important topic in theoretical physics, which is studied in the physical and physico-mathematical departments of universities. Therefore, lecturers are faced with the urgent task of not only providing students with information about the subject, but also to help them master the material at a deep qualitative level, by presenting the specific features of general approaches to the statement and the solution of problems in theoretical physics. One of the ways to study field theory is the practical one, where the students can deepen their knowledge of the theoretical material and develop problem-solving skills. This book includes a concise theoretical summary of the main branches of field theory and electrodynamics, worked examples, and some problems for the student to solve.

The book is written for students of theoretical and applied physics, and corresponds to the curricula of the theoretical courses "Field theory" and "Electrodynamics" for physics undergraduates. It can also be useful for students of other disciplines, in particular, those in which physics is one of the base subjects.

Introduction

The active influence of science on the production and the society requires that an up-to-date student, having graduated from a university, be a comprehensively developed, creatively thinking, and socially active person. He/she should be gifted in scientific-technical creative work, use the achievements of science in practical activity, participate in scientific-research work, and be guided by the principles of scientific organization of work and management. The training of such specialists at a university must be executed on the basis of the development of creative forms of education for special and general disciplines.

Physics occupies a prominent place in the fundamental sciences near mathematics, chemistry, etc. Of high importance is the combination of the fundamental character of education of students and the ability to efficiently apply the results of physical studies to scientific-technical progress. This discipline belongs to the class of subjects "Theoretical physics" and is based on the knowledge of physics and mathematics according to the secondary-school program and the courses "Mathematical analysis," "Analytic geometry," "Electricity and magnetism," "Classical mechanics," etc. Knowledge obtained by students from the course "Electrodynamics" is used in the courses "Quantum mechanics," "Theory of solids," etc. While studying the course "Field theory," the students should: (1) master the basic principles and laws of the certain branches of physics and their mathematical description, make the acquaintance of main physical phenomena and methods of their observation and experimental studies; learn to properly reproduce the physical ideas, quantitatively formulate and solve physical problems, evaluate the order of physical quantities; clearly see the limits of application of physical models and the theory; (2) master the comprehension of philosophical and methodological problems of modern science; (3) properly imagine the role of each branch of physics in scientific-technical progress and develop the interest and the skill to solve the scientific-technical and applied tasks. In addition, the students of physico-mathematical faculties study such theoretical engineering courses as "Theoretical mechanics," "Theoretical foundations of electrotechnique," "Computational technique," etc. Therefore, special attention in the courses "Electrodynamics" and "Field theory" is paid to physical laws and to the relation of physical notions to experimental data and certain methods of measurements. During the teaching of the course "Field theory," the accent is focused on two closely connected aspects: representation of the physical essence of phenomena and analysis of analytic relations describing those phenomena.

In view of the variety of the forms of matter and motion studied by physics, the lecturing of the course takes the technical orientation of the faculty into account to a certain degree. At the same time, the principal role under conditions of the fast scientific-technical progress is assigned to the theoretical scientific-technical level of a specialist, who should deal successfully with the newest branches of science and technique. By mastering the course "Field theory", the students of physico-mathematical faculties must know the fundamental laws of this branch of physics and methods of their studies with complete comprehension and apply this knowledge to the consideration of separate phenomena, by describing their physical essence with analytic relations, and to the study of other engineering and specialized disciplines. This book helps to develop an independent approach to any physical problem, to learn to understand, think and reflect.

It is worth noting that, in the process of writing of this manual, we have used the classical sources [1, 2], which became a foundation for the presentation of theoretical positions, whereas the typical methods of solution of the problems are based on materials given in [3–12].

Contents

Section 1

Vector and Tensor Analyses

1.1. Vector and tensor algebras.
Transformation of vectors and tensors

Scalar (invariant) in three-dimensional space is a quantity that is invariant under rotations (inversion) of a coordinate system.

Vector in three-dimensional space is a collection of three quantities that are transformed under rotations of a coordinate system by the formulas

$$A'_\alpha = \sum_{\beta=1}^{3} a_{\alpha\beta} A_\beta \qquad (1.1)$$

or, according to the rules of tensor analysis,

$$A'_\alpha = a_{\alpha\beta} A_\beta$$

(where the summation over the repeated indices is assumed). Here, A'_α is the projection of the vector on the α-th axis of the rotated coordinate system; $a_{\alpha\beta}$ is the coefficient of transformation, which is the cosine of the angle between the β-th axis of the initial coordinate system and the α-th axis of the rotated one; A_β is the projection of the vector on the β-th axis of the initial system.

Vectors can be written in terms of contravariant (A^α) or covariant (A_α) coordinates. The square of a three-dimensional vector is the quantity

$$\sum_{\alpha=1}^{3} A^\alpha A_\alpha = A^1 A_1 + A^2 A_2 + A^3 A_3.$$

1

The scalar product of two vectors is defined as

$$A^\alpha B_\alpha = A^1 B_1 + A^2 B_2 + A^3 B_3.$$

Second-rank tensor in three-dimensional space is the nine-component quantity $T_{\alpha\beta}(\alpha, \beta = 1, 2, 3)$, which is transformed under rotations of a coordinate system in the following way:

$$T'_{\alpha\beta} = \alpha_{\alpha\lambda}\alpha_{\beta\mu}T_{\lambda\mu}$$

(as above, the sum over λ and μ is assumed).

Analogously, a *third-rank tensor* in a three-dimensional space is defined by the law of transformation:

$$T'_{\alpha\beta\gamma} = \alpha_{\alpha\lambda}\alpha_{\beta\mu}\alpha_{\gamma\nu}T_{\lambda\mu\nu}.$$

The tensors of higher ranks are defined analogously.

The vector quantities under inversion of a coordinate system can be transformed in two ways. Those vectors, whose components at the inversion of coordinates change sign (transformation $x' = -x$, $y' = -y$, $z' = -z$), are called *polar* vectors or simply vectors. The vectors, whose components do not change sign under inversion of a coordinate system, are called *pseudovectors* or *axial* vectors. (The difference between the covariant and contravariant components of vectors and tensors is not essential for the problems considered in this section).

An example of an axial vector is the vector product of two polar vectors. Analogously, the tensor of the s-th rank is called simply a *tensor*, if its components are transformed under inversion as the product of s coordinates, i.e., they are multiplied by $(-1)^s$, and a *pseudotensor*, if its components are multiplied by $(-1)^{s+1}$.

The table of the coefficients of a transformation

$$\hat{\alpha} = \begin{pmatrix} \alpha_{11} & \alpha_{12} & \alpha_{13} \\ \alpha_{21} & \alpha_{22} & \alpha_{23} \\ \alpha_{31} & \alpha_{32} & \alpha_{33} \end{pmatrix}$$

is called the *matrix of transformation*. In what follows, we will consider the *determinant* of this matrix

$$\det \hat{\alpha} = \begin{vmatrix} \alpha_{11} & \alpha_{12} & \alpha_{13} \\ \alpha_{21} & \alpha_{22} & \alpha_{23} \\ \alpha_{31} & \alpha_{32} & \alpha_{33} \end{vmatrix}.$$

The *sum of two matrices* $\hat{\alpha} + \hat{\beta}$ is the matrix $\hat{\gamma}$, whose elements are equal to the sums of corresponding elements of the matrices-summands:

$$\gamma_{\alpha\beta} = \alpha_{\alpha\beta} + \beta_{\alpha\beta}.$$

The *product of two matrices* $\hat{\alpha}\hat{\beta}$ is the matrix $\hat{\gamma}$, whose elements are formed from elements of the matrices $\alpha_{\alpha\beta}$ and $\beta_{\alpha\beta}$, which are multiplied by the rule:

$$\gamma_{\alpha\beta} = \alpha_{\alpha\gamma}\beta_{\gamma\beta}.$$

The matrix $\hat{\gamma}$ describes the transformation that is carried out by two subsequent transformations: first, that corresponding to the matrix $\hat{\beta}$, and then to the matrix $\hat{\alpha}$.

The *identity matrix* is a matrix of the type:

$$\hat{1} = \begin{pmatrix} 1 & 0 & 0 \\ 0 & 1 & 0 \\ 0 & 0 & 1 \end{pmatrix}.$$

It describes the identity transformation $(A'_\alpha = A_\alpha)$. Its elements are denoted by the symbol $\delta_{\alpha\beta}$:

$$\delta_{\alpha\beta} = \begin{cases} 1, & \text{if } \alpha = \beta; \\ 0, & \text{if } \alpha \neq \beta. \end{cases}$$

A matrix of the type

$$\hat{\alpha} = \begin{pmatrix} \alpha_1 & 0 & 0 \\ 0 & \alpha_2 & 0 \\ 0 & 0 & \alpha_3 \end{pmatrix}$$

is called a *diagonal matrix*.

If the elements of a matrix satisfy the condition

$$\alpha_{\alpha\beta}\alpha_{\alpha\gamma} = \delta_{\beta\gamma},$$

then it is called *orthogonal*.

The matrix $\hat{\alpha}^{-1}$ satisfying the conditions

$$\hat{\alpha}\hat{\alpha}^{-1} = \hat{\alpha}^{-1}\hat{\alpha} = \hat{1}$$

is called the *inverse* to the matrix $\hat{\alpha}$.

The former describes the inverse transformation: i.e., if $A'_\alpha = \alpha_{\alpha\beta}A_\beta$, then $A_\alpha = \alpha_{\alpha\beta}^{-1}A'_\beta$.

The matrix $\hat{\tilde{\alpha}}$ formed from $\hat{\alpha}$ by the substitution of columns for rows is called *transposed*:

$$\hat{\tilde{\alpha}} = \begin{pmatrix} \alpha_{11} & \alpha_{21} & \alpha_{31} \\ \alpha_{12} & \alpha_{22} & \alpha_{32} \\ \alpha_{13} & \alpha_{23} & \alpha_{33} \end{pmatrix}, \quad \tilde{\alpha}_{\alpha\beta} = \alpha_{\beta\alpha}.$$

Matrices and tensors can be symmetric and antisymmetric.

The second-rank tensor $S_{\alpha\beta}$ is *symmetric* if its components remain invariable under permutation of indices: $S_{\alpha\beta} = S_{\beta\alpha}$.

A tensor $A_{\alpha\beta}$, whose components change sign under permutation of indices, is called *antisymmetric*: $A_{\alpha\beta} = -A_{\beta\alpha}$.

The Kronecker delta $\delta_{\alpha\beta}$ is the *unit second-rank tensor*, whose components are the same in all coordinate systems, like the tensor $e_{\alpha\beta\gamma}$ which is the *absolutely antisymmetric unit third-rank pseudotensor*. The collection of components $e_{\alpha\beta\gamma}$ has the following properties: under permutation of two any indices, the nonzero component of $e_{\alpha\beta\gamma}$ changes its sign, and $e_{123} = 1$. Of the 27 values of $e_{\alpha\beta\gamma}$, only six values are nonzero. Many have at least two identical indices and are zero due to the antisymmetry ($e_{\alpha\alpha\gamma} = -e_{\alpha\alpha\gamma} = 0$). The nonzero components are as follows:

$$e_{123} = e_{231} = e_{312} = -e_{321} = -e_{213} = -e_{132} = 1.$$

The tensors $\delta_{\alpha\beta}$ and $e_{\alpha\beta\gamma}$ play important roles in vector and tensor algebras.

* * *

Example 1.1. Let, in all Cartesian coordinate systems, the collection of three quantities a_α ($\alpha = 1, 2, 3$) be given, and let $a_\alpha b_\alpha = $ inv at rotations and reflections. Prove that if b_α is a vector (pseudovector), then a_α is also a vector (pseudovector).

Solution. Since b_α are components of a vector, they are transformed under rotation of the coordinate system by formulas (1.1): $b'_\alpha = \alpha_{\alpha\beta} b_\beta$. Substituting b'_α in the equality $a'_\alpha b'_\alpha = $ inv and comparing with $a_\alpha b_\alpha = $ inv, we get $a_\beta = \alpha_{\alpha\beta} a'_\alpha$, i.e., a_β are transformed under rotations as components of a vector. Since the invariant does not change sign under reflections, the components of a_α and b_α should simultaneously change sign (polar vectors) or should not (pseudovectors).

Example 1.2. Construct the matrices of the transformation of basis unit vectors for the transition from Cartesian coordinates to cylindrical ones and those of the inverse transformation.

Solution. The *spherical* $\vec{r} = (R, \vartheta, \alpha)$ (Fig. 1.1) and *cylindrical* $\vec{r} = (r, \alpha, z)$ (Fig. 1.2) coordinate systems are widely used in the solution of problems of theoretical physics along with the Cartesian coordinate system, where the components of a radius-vector are given as $\vec{r} = (x, y, z)$.

Under the transition from the Cartesian to spherical coordinate system ($\vec{e}_\alpha \to \vec{e}'_\alpha$) by the formulas $\vec{e}'_\alpha = \alpha_{\alpha\beta} \vec{e}_\alpha$, we pass from the

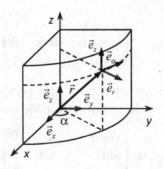

Fig. 1.1

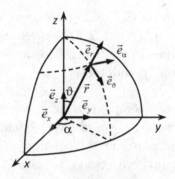

Fig. 1.2

unit vectors $\vec{e}_x, \vec{e}_y, \vec{e}_z$ to the unit vectors $\vec{e}_r, \vec{e}_\alpha, \vec{e}_z$. In this case, the coefficients $\alpha_{\alpha\beta}$ are the projections of the new unit vectors onto the old ones:

$$\alpha_{\alpha\beta} = \begin{pmatrix} \alpha_{rx} & \alpha_{ry} & \alpha_{rz} \\ \alpha_{\alpha x} & \alpha_{\alpha y} & \alpha_{\alpha z} \\ \alpha_{zx} & \alpha_{zy} & \alpha_{zz} \end{pmatrix}.$$

For example, α_{rx} is the coefficient which is determined by projecting $\vec{e}_r$ onto $\vec{e}_x$: the angle between these unit vectors is α. Therefore, in order to project $\vec{e}_r$ onto $\vec{e}_x$, we need to multiply $\vec{e}_r$ by $\cos \alpha$. Respectively, $\alpha_{rx} = \cos \alpha$.

By determining in this way the projections of some unit vectors onto other ones, we get the matrix of the transformation from Cartesian coordinates to cylindrical ones:

$$\alpha_{\alpha\beta} = \begin{pmatrix} \cos \alpha & \sin \alpha & 0 \\ -\sin \alpha & \cos \alpha & 0 \\ 0 & 0 & 1 \end{pmatrix}.$$

By projecting analogously the old unit vectors onto the new ones, we obtain the matrix of the inverse transformation:

$$\alpha_{\alpha\beta}^{-1} = \begin{pmatrix} \cos \alpha & -\sin \alpha & 0 \\ \sin \alpha & \cos \alpha & 0 \\ 0 & 0 & 1 \end{pmatrix}.$$

Example 1.3. Let $\vec{n}$ be a unit vector, whose direction in space are equiprobable. Find the mean values of its components and their products: $\overline{n_\alpha}$, $\overline{n_\alpha n_\beta}$, $\overline{n_\alpha n_\beta n_\gamma}$, $\overline{n_\alpha n_\beta n_\gamma n_\delta}$.

Solution. The mean values are equal to the integrals:

$$\overline{n_\alpha} = \frac{1}{4\pi} \int n_\alpha \, d\Omega, \quad \overline{n_\alpha n_\beta} = \frac{1}{4\pi} \int n_\alpha n_\beta \, d\Omega.$$

However, instead of the direct calculation of these integrals, it is more convenient to use the transformational properties of the analyzed quantities. It is obvious that the quantities $\overline{n_\alpha}$, $\overline{n_\alpha n_\beta}$, etc. are tensors, respectively, of the first, second, etc. ranks. It follows from their definition that these quantities must be the same in any reference system. Therefore, they can be given in terms of such tensors, whose components are independent of the choice of the reference system.

Let us consider $\overline{n_\alpha}$. Since there is no vector, except for the zero one, whose components are independent of the reference system, we have $\overline{n_\alpha} = 0$.

The tensor $\overline{n_\alpha n_\beta}$ must be presented in terms of a symmetric second-rank tensor, whose components are the same in all reference systems. The only such tensor is $\delta_{\alpha\beta}$. Therefore, we can write $\overline{n_\alpha n_\beta} = \lambda \delta_{\alpha\beta}$. To determine λ, the tensor should be convolved on the pair of symbols: $\overline{n_\alpha n_\alpha} = n^2 = 1 = 3\lambda$, $\lambda = 1/3$. Analogously, we get

$$\overline{n_\alpha n_\beta n_\gamma n_\delta} = 1/15(\delta_{\alpha\beta}\delta_{\gamma\delta} + \delta_{\alpha\gamma}\delta_{\beta\delta} + \delta_{\alpha\delta}\delta_{\beta\gamma}).$$

Example 1.4. Find the components of the tensor $\varepsilon_{\alpha\beta}^{-1}$, inverse to the tensor $\varepsilon_{\alpha\beta}$. Consider, in particular, the case where $\varepsilon_{\alpha\beta}$ is a symmetric tensor given in the principal axes.

Answer. The tensor inverse to the given one satisfies the relation

$$\varepsilon_{\alpha\beta}\varepsilon_{\beta\gamma}^{-1} = \delta_{\alpha\gamma}. \tag{1.2}$$

It is the algebraic equation for components of the inverse tensor $\varepsilon_{\alpha\beta}^{-1}$. Its solution takes the form

$$\varepsilon_{\alpha\beta}^{-1} = \frac{\Delta_{\beta\alpha}}{|\varepsilon|}, \tag{1.3}$$

where $\Delta_{\beta\alpha}$ is the cofactor of an element $\varepsilon_{\alpha\beta}$.

Formula (1.3) implies that the necessary condition of the existence of the inverse tensor is $|\varepsilon| \neq 0$. In view of the well-known property $\Delta_{\beta\alpha}\varepsilon_{\beta\gamma} = \delta_{\alpha\gamma}|\varepsilon|$, we see with regard for expression (1.2) that the inverse tensor satisfies also the condition

$$\varepsilon_{\alpha\beta}^{-1}\varepsilon_{\beta\gamma} = \delta_{\alpha\gamma}.$$

If $\varepsilon_{\alpha\beta}$ is a symmetric tensor given in the principal axes, $\varepsilon_{\alpha\beta} = \varepsilon^{(\alpha)}\delta_{\alpha\beta}$ (no summation over α), then $\varepsilon_{\alpha\beta}^{-1} = \frac{1}{\varepsilon^{(\alpha)}}\delta_{\alpha\beta}$.

Problems

1.1. Two directions $\vec{n}$ and $\vec{n}'$ are defined in a spherical coordinate system by the angles ϑ, α and ϑ', α, respectively. Find the cosine of the angle between them.

    Answer. $\cos\vartheta = n\,n' = \cos\vartheta\cos\vartheta' + \sin\vartheta\sin\vartheta'\cos(\alpha - \alpha')$.

1.2. Prove the identities:

    (a) $[\vec{A} \times \vec{B}] \cdot [\vec{C} \times \vec{D}] = (\vec{A} \cdot \vec{C})(\vec{B} \cdot \vec{D}) - (\vec{A} \cdot \vec{D})(\vec{B} \cdot \vec{C})$;

    (b) $[\vec{A} \times \vec{B}] \times [\vec{C} \times \vec{D}] = (\vec{A} \cdot [\vec{B} \times \vec{D}])\vec{C} - (\vec{A} \cdot [\vec{B} \times \vec{C}])\vec{D}$

$$= (\vec{A} \cdot [\vec{C} \times \vec{D}])\vec{B} - (\vec{B} \cdot [\vec{C} \times \vec{D}])\vec{A}.$$

1.3. Prove that if $a_\alpha = T_{\alpha\beta}b_\beta$ in any coordinate system, $T_{\alpha\beta}$ is a second-rank tensor, and b_β is a vector, then a_α is also a vector.

1.4. Prove that $\partial a_\alpha / \partial x_\beta$ is a second-rank tensor.

1.5. Prove that if $T_{\alpha\beta}$ is a second-rank tensor, and if $P_{\alpha\beta}$ is a second-rank pseudotensor, then $T_{\alpha\beta}P_{\alpha\beta}$ is a pseudoscalar.

1.6. Show that if the tensor $S_{\alpha\beta}$ is symmetric, and the tensor $A_{\alpha\beta}$ is antisymmetric, then $A_{\alpha\beta}S_{\alpha\beta} = 0$.

1.7. Prove that the sum of diagonal components of a second-rank tensor is an invariant.

1.8. Let the components of the vector $\vec{a}$ in all coordinate systems be linearly presented in terms of components of the vector $\vec{b}$: $a_\alpha = \varepsilon_{\alpha\beta} b_\beta$. Prove that the collection of quantities $\varepsilon_{\alpha\beta}$ is a second-rank tensor.

1.9. Show that the collection of quantities $A_{\alpha\beta\gamma} B_{\alpha\beta}$, where $A_{\alpha\beta\gamma}$ is a third-rank tensor, and $B_{\alpha\beta}$ is a second-rank tensor, is a vector.

1.10. Find the law of transformation of the collection of volume integrals $T_{\alpha\beta} = \int x_\alpha x_\beta \, dV$ under spatial rotations and reflections (x_α and x_β are Cartesian coordinates).

Answer. $T_{\alpha\beta}$ form a second-rank tensor.

1.11. Construct the matrices of transformation of basis unit vectors: under the transition from the Cartesian coordinates to spherical ones and, conversely, from the spherical coordinates to Cartesian ones; under the transition from the Cartesian coordinates to cylindrical ones and, conversely, from the cylindrical coordinates to Cartesian ones.

1.12. In all Cartesian coordinate systems, let us set a collection of quantities $e_{\alpha\beta\gamma}$ with the following properties: $e_{\alpha\beta\gamma}$ changes its sign under the permutation of any two indices, and $e_{123} = 1$.

Show that the collection $e_{\alpha\beta\gamma}$ forms a third-rank pseudotensor (totally antisymmetric unit third-rank pseudotensor).

Answer. Only 6 quantities of the 27 are nonzero. The remaining quantities have at least two identical indices and are transformed to zero due to the antisymmetry ($e_{\alpha\alpha\beta} = -e_{\alpha\alpha\beta} = 0$). The nonzero components are (see page 4)

$$e_{123} = e_{231} = e_{312} = -e_{321} = -e_{213} = -e_{132} = 1.$$

Let us consider the expression $\alpha_{1\alpha}\alpha_{2\beta}\alpha_{3\gamma} e_{\alpha\beta\gamma}$. Remembering the definition of a third-order determinant and using the definition of $e_{\alpha\beta\gamma}$, we write this expression in the form $\alpha_{1\alpha}\alpha_{2\beta}\alpha_{3\gamma} e_{\alpha\beta\gamma} = |\hat{\alpha}| = +1 = e'_{\alpha\beta\gamma}$. Let us transpose two indices on the left-hand side, for example, 1 and 2. We get

$$\alpha_{2\alpha}\alpha_{1\beta}\alpha_{3\gamma} e_{\alpha\beta\gamma} = -\alpha_{1\beta}\alpha_{2\alpha}\alpha_{3\gamma} e_{\alpha\beta\gamma} = -e'_{123} = e'_{213} \cdots$$

It is seen from these relations that $e_{\alpha\beta\gamma}$ is transformed under rotations as a third-rank tensor. The quantities $e_{\alpha\beta\gamma}$ are invariant under the transformations. Therefore, their collection forms a third-rank axial tensor. It has a valuable property: its components are the same in all coordinate systems.

1.13. Prove that the components of a second-rank antisymmetric tensor under a rotation of coordinates are transformed as the components of a vector.

Answer. Let us write the tensor $A_{\alpha\beta}$ in the form of a table:

$$A_{\alpha\beta} = \begin{pmatrix} 0 & A_{12} & -A_{31} \\ -A_{21} & 0 & A_{23} \\ A_{31} & -A_{23} & 0 \end{pmatrix}.$$

Denote: $A_{23} = A_1$, $A_{31} = A_2$, $A_{12} = A_3$. These three relations can be rewritten as $A_\alpha = \frac{1}{2}e_{\alpha\beta\gamma}A_{\beta\gamma}$, where $e_{\alpha\beta\gamma}$ is the totally antisymmetric third-rank unit tensor. But since $e_{\alpha\beta\gamma}$ is a third-rank tensor, and $A_{\beta\gamma}$ is a second-rank tensor, the quantities A_α ($\alpha = 1, 2, 3$) form a vector; A_α is called the vector dual to the tensor $A_{\alpha\beta}$.

1.14. To write the formulas for components of the vector product of two vectors and the curl of a vector with the help of the unit antisymmetric tensor $e_{\alpha\beta\gamma}$. Determine how these quantities are transformed under rotations and reflections.

Answer. $[\vec{A} \times \vec{B}]_\alpha = -e_{\alpha\beta\gamma}A_\beta B_\gamma$, $\text{rot}_\alpha \vec{A} = e_{\alpha\beta\gamma}\frac{\partial A_\gamma}{\partial x_\beta}$, $[\vec{A} \times \vec{B}]$ and rot $\vec{A}$ can be considered as antisymmetric second-rank tensors or as vectors dual to them, whose components do not change their signs under a transformation (pseudovectors).

1.15. Prove the equalities:

(a) $e_{\alpha\beta\gamma}e_{\gamma\delta\varepsilon} = \delta_{\alpha\delta}\delta_{\beta\varepsilon} - \delta_{\alpha\varepsilon}\delta_{\beta\delta}$;

(b) $e_{\alpha\beta\gamma}e_{\beta\gamma\delta} = 2\delta_{\alpha\delta}$.

1.16. Show that $T_{\alpha\beta}a_\alpha b_\beta - T_{\alpha\beta}a_\beta b_\alpha = 2\vec{\omega} \cdot [\vec{a} \times \vec{b}]$, where $T_{\alpha\beta}$ is any second-rank tensor, $\vec{a}$ and $\vec{b}$ are vectors, and $\vec{\omega}$ is the vector equivalent to the antisymmetric part of $T_{\alpha\beta}$.

1.17. Transform the product $(\vec{a} \cdot [\vec{b} \times \vec{c}])(\vec{a}' \cdot [\vec{b}' \times \vec{c}'])$ into the sum of terms that contain only the scalar products of vectors.

Hint. Apply the theorem of multiplication of determinants or use the third-rank pseudotensor $e_{\alpha\beta\gamma}$.

Answer.

$$(\vec{a} \cdot \vec{a}')(\vec{b} \cdot \vec{b}')(\vec{c} \cdot \vec{c}') + (\vec{a} \cdot \vec{b}')(\vec{b} \cdot \vec{c}')(\vec{c} \cdot \vec{a}')$$
$$+ (\vec{b} \cdot \vec{a}')(\vec{c} \cdot \vec{b}')(\vec{a} \cdot \vec{c}') - (\vec{a} \cdot \vec{c}')(\vec{c} \cdot \vec{a}')(\vec{b} \cdot \vec{b}')$$
$$- (\vec{a} \cdot \vec{b}')(\vec{b} \cdot \vec{a}')(\vec{c} \cdot \vec{c}') - (\vec{b} \cdot \vec{c}')(\vec{c} \cdot \vec{b}')(\vec{a} \cdot \vec{a}').$$

1.18. Find the values of the following expressions averaged over all directions: $\overline{(\vec{a} \cdot \vec{n})^2}$, $\overline{(\vec{a} \cdot \vec{n})(\vec{b} \cdot \vec{n})}$, $\overline{(\vec{a} \cdot \vec{n})\vec{n}}$, $\overline{[\vec{a} \times \vec{n}]^2}$, $\overline{[\vec{a} \times \vec{n}][\vec{b} \times \vec{n}]}$, $\overline{(\vec{a} \cdot \vec{n})(\vec{b} \cdot \vec{n})(\vec{c} \cdot \vec{n})(\vec{d} \cdot \vec{n})}$, if $\vec{n}$ is a unit vector, whose all directions are equiprobable, and $\vec{a}$, $\vec{b}$, $\vec{c}$, and $\vec{d}$ are constant vectors.

Hint. Use the results of Problem 1.18.

Answer.

$$a^2/3, \quad (\vec{a} \cdot \vec{b})/3, \quad \vec{a}/3, \quad 2/3a^2, \quad 2/3(\vec{a} \cdot \vec{b});$$
$$1/15[(\vec{a} \cdot \vec{b})(\vec{c} \cdot \vec{d}) + (\vec{a} \cdot \vec{c})(\vec{b} \cdot \vec{d}) + (\vec{a} \cdot \vec{d})(\vec{b} \cdot \vec{c})].$$

1.2. Vector analysis

In an arbitrary orthogonal coordinate system q_1, q_2, q_3, the square of the length element is given by the formula

$$dl^2 = h_1^2 dq_1^2 + h_2^2 dq_2^2 + h_3^2 dq_3^2,$$

and the volume element

$$dV = h_1 h_2 h_3 dq_1 dq_2 dq_3,$$

where

$$h_i = \sqrt{\left(\frac{\partial x}{\partial q_i}\right)^2 + \left(\frac{\partial y}{\partial q_i}\right)^2 + \left(\frac{\partial z}{\partial q_i}\right)^2}$$

are functions of the coordinates (Lamé coefficients).

The differential operations are written as follows:

$$\operatorname{div} \vec{A} = \frac{1}{h_1 h_2 h_3} \left[\frac{\partial}{\partial q_1}(h_2 h_3 A_1) + \frac{\partial}{\partial q_2}(h_1 h_3 A_2) + \frac{\partial}{\partial q_3}(h_1 h_2 A_3) \right];$$

$$\operatorname{rot} \vec{A} = \begin{vmatrix} \dfrac{\vec{e}_1}{h_2 h_3} & \dfrac{\vec{e}_2}{h_1 h_3} & \dfrac{\vec{e}_3}{h_1 h_2} \\[2mm] \dfrac{\partial}{\partial q_1} & \dfrac{\partial}{\partial q_2} & \dfrac{\partial}{\partial q_3} \\[2mm] h_1 A_1 & h_2 A_2 & h_3 A_3 \end{vmatrix}; \quad (\operatorname{grad} \varphi)_i = \frac{1}{h_i}\frac{\partial \varphi}{\partial q_i};$$

$$\Delta \varphi = \frac{1}{h_1 h_2 h_3} \left[\frac{\partial}{\partial q_1}\left(\frac{h_2 h_3}{h_1}\frac{\partial \varphi}{\partial q_1} \right) + \frac{\partial}{\partial q_2}\left(\frac{h_1 h_3}{h_2}\frac{\partial \varphi}{\partial q_2} \right) \right.$$
$$\left. + \frac{\partial}{\partial q_3}\left(\frac{h_1 h_2}{h_3}\frac{\partial \varphi}{\partial q_3} \right) \right].$$

In the formula for rot $\vec{A}$, the differential operators $\frac{\partial}{\partial q_i}$ act on the elements of the last row of the determinant. In a *spherical* coordinate system:

$$x = r \sin \vartheta \cos \alpha, \quad y = r \sin \vartheta \sin \alpha, \quad z = r \cos \vartheta;$$
$$h_r = 1, \quad h_\vartheta = r, \quad h_\alpha = r \sin \vartheta;$$
$$\operatorname{grad} \varphi = \vec{e}_r \frac{\partial \varphi}{\partial r} + \frac{\vec{e}_\vartheta}{r}\frac{\partial \varphi}{\partial \vartheta} + \frac{\vec{e}_\alpha}{r \sin \alpha}\frac{\partial \varphi}{\partial \alpha};$$
$$\operatorname{div} \vec{A} = \frac{1}{r^2}\frac{\partial}{\partial r}(r^2 A_r) + \frac{1}{r \sin \vartheta}\frac{\partial}{\partial \vartheta}(\sin \vartheta\, A_\vartheta) + \frac{1}{r \sin \vartheta}\frac{\partial A_\alpha}{\partial \alpha};$$
$$(\operatorname{rot} \vec{A})_r = \frac{1}{r \sin \vartheta} \left[\frac{\partial}{\partial \vartheta}(\sin \vartheta\, A_\alpha) - \frac{\partial A_\vartheta}{\partial \alpha} \right];$$
$$(\operatorname{rot} \vec{A})_\vartheta = \frac{1}{r \sin \vartheta}\frac{\partial A_r}{\partial \alpha} - \frac{1}{r}\frac{\partial(r A_\alpha)}{\partial r};$$
$$(\operatorname{rot} \vec{A})_\alpha = \frac{1}{r}\frac{\partial(r A_\vartheta)}{\partial r} - \frac{1}{r}\frac{\partial A_r}{\partial \vartheta};$$
$$\Delta \varphi = \frac{1}{r^2}\frac{\partial}{\partial r}\left(r^2 \frac{\partial \varphi}{\partial r} \right) + \frac{1}{r^2 \sin \vartheta}\frac{\partial}{\partial \vartheta}\left(\sin \vartheta \frac{\partial \varphi}{\partial \vartheta} \right) + \frac{1}{r^2 \sin^2 \vartheta}\frac{\partial^2 \varphi}{\partial \alpha^2}.$$

In a *cylindrical* coordinate system:

$$x = r \cos \alpha, \quad y = r \sin \alpha, \quad z = z;$$

$$h_r = 1, \quad h_\alpha = r, \quad h_z = 1;$$

$$\operatorname{grad} \varphi = \vec{e}_r \frac{\partial \varphi}{\partial r} + \frac{\vec{e}_\vartheta}{r} \frac{\partial \varphi}{\partial \alpha} + \vec{e}_z \frac{\partial \varphi}{\partial z};$$

$$\operatorname{div} \vec{A} = \frac{1}{r} \frac{\partial}{\partial r}(r A_r) + \frac{1}{r} \frac{\partial A_\alpha}{\partial \alpha} + \frac{\partial A_z}{\partial z};$$

$$(\operatorname{rot} \vec{A})_r = \frac{1}{r} \frac{\partial A_z}{\partial \alpha} - \frac{\partial A_\alpha}{\partial z}; \quad (\operatorname{rot} \vec{A})_\alpha = \frac{\partial A_r}{\partial z} - \frac{\partial A_z}{\partial r};$$

$$(\operatorname{rot} \vec{A})_z = \frac{1}{r} \frac{\partial}{\partial r}(r A_\alpha) - \frac{1}{r} \frac{\partial A_r}{\partial \alpha};$$

$$\Delta \varphi = \frac{1}{r} \frac{\partial}{\partial r}\left(r \frac{\partial \varphi}{\partial r}\right) + \frac{1}{r^2} \frac{\partial^2 \varphi}{\partial \alpha^2} + \frac{\partial^2 \varphi}{\partial z^2}.$$

For any $\vec{A}$ and φ, the following identities hold:

$$\operatorname{rot} \operatorname{grad} \varphi \equiv 0, \quad \operatorname{div} \operatorname{rot} \varphi \equiv 0, \quad \operatorname{div} \operatorname{grad} \varphi \equiv \Delta \varphi.$$

The below-presented basic integral theorems connect volume, surface, and contour integrals.

Ostrogradskii–Gauss theorem:

$$\int_V \operatorname{div} \vec{A} dV = \oint_S \vec{A} \cdot d\vec{S}, \tag{1.4}$$

where V is some volume; S is the closed surface that bounds this volume.

Stokes theorem:

$$\oint_l \vec{A} \cdot d\vec{l} = \int_S \operatorname{rot} \vec{A} \cdot d\vec{S}, \tag{1.5}$$

where l is a closed contour; and S is any surface that relies on this contour.

In formulas (1.4) and (1.5), the vector $\vec{A}$ must be a differentiable function of the coordinates.

$$* * *$$

Example 1.5. Using the Ostrogradskii–Gauss theorem, calculate the integral

$$\vec{I} = \oint \vec{r}(\vec{a} \cdot \vec{n}) \, dS,$$

if the volume enveloped by the closed surface is equal to V; $\vec{a}$ is a constant vector, and $\vec{n}$ is a unit vector normal to the surface.

Solution. Here, like in a number of other cases, it is convenient to consider the scalar product of the integral by any constant vector $\vec{c}$:

$$\vec{c} \cdot \oint \vec{r}(\vec{a} \cdot \vec{n}) dS = \oint (\vec{c} \cdot \vec{r}) a_n dS = \int \operatorname{div}((\vec{c} \cdot \vec{r})\vec{a}) dV$$

$$= (\vec{a} \cdot \vec{c}) \int dV = (\vec{a} \cdot \vec{c}) V.$$

Since $\vec{c}$ is a constant vector, this implies that $\vec{I} = \vec{a} V$.

Problems

1.19. Calculate $\operatorname{div} \vec{r}$, $\operatorname{rot} \vec{r}$, $\operatorname{grad}(\vec{p} \cdot \vec{r})$, $\operatorname{grad} \frac{\vec{p}\vec{r}}{r^3}$, and $(\vec{p} \cdot \nabla)\vec{r}$, where $\vec{r}$ is a radius-vector; and $\vec{p}$ is a constant vector.

1.20. Find the divergences and the curls of the vectors $(\vec{a} \cdot \vec{r})\vec{b}$, $(\vec{a} \cdot \vec{r})\vec{r}$, $[\vec{a} \times \vec{r}]$, and $\vec{r} \times [\vec{a} \times \vec{r}]$, where $\vec{a}$ and $\vec{b}$ are constant vectors.

Answer.

$$\operatorname{div}(\vec{a} \cdot \vec{r})\vec{b} = \vec{a} \cdot \vec{b}, \quad \operatorname{rot}(\vec{a} \cdot \vec{r})\vec{b} = \vec{a} \times \vec{b}, \quad \operatorname{div}(\vec{a} \cdot \vec{r})\vec{r} = 4(\vec{a} \cdot \vec{r});$$

$$\operatorname{rot}(\vec{a} \cdot \vec{r})\vec{r} = \vec{a} \times \vec{r}, \quad \operatorname{div}[\vec{a} \times \vec{r}]\vec{r} = 0, \quad \operatorname{rot}[\vec{a} \times \vec{r}] = 2\vec{a};$$

$$\operatorname{div} \vec{r} \times [\vec{a} \times \vec{r}] = -2(\vec{a} \cdot \vec{r}), \quad \operatorname{rot} \vec{r} \times [\vec{a} \times \vec{r}] = 3[\vec{r} \times \vec{a}].$$

1.21. Calculate $\operatorname{grad} \varphi(r)$, $\operatorname{div} \varphi(r)\vec{r}$, $\operatorname{rot}\varphi(r)\vec{r}$, $(\vec{p} \cdot \nabla)\varphi(r)\vec{r}$. The function $\varphi(r)$ depends only on the modulus of the radius-vector.

Answer. $\operatorname{grad} \vec{A}(r) \cdot \vec{B}(r)$, $\operatorname{grad} \varphi(r) = \frac{\vec{r}}{r}\varphi'$, $\operatorname{div} \varphi(r)\vec{r} = 3\varphi + r\varphi'$; $\operatorname{rot}\varphi(r)\vec{r} = 0$, $(\vec{p} \cdot \nabla)\varphi(r)\vec{r} = \vec{p}\varphi + \frac{\vec{r}(\vec{p}\cdot\vec{r})}{r}\varphi'$.

1.22. Calculate $\operatorname{grad} \vec{A}(r) \cdot \vec{r}$, $\operatorname{div} \varphi(r) \cdot \vec{A}(r)$, $\operatorname{rot}\varphi(r) \cdot \vec{A}(r)$, $(\vec{p} \cdot \nabla)$ $\varphi(r)\vec{A}(r)$.

Answer.

$$\operatorname{grad}\vec{A}(r) \cdot \vec{r} = \vec{A} + \frac{\vec{r}}{r}(\vec{r} \cdot \vec{A}');$$

$$\operatorname{grad}\vec{A}(r) \cdot \vec{B}(r) = \frac{\vec{r}}{r}(\vec{A}' \cdot \vec{B} + \vec{A} \cdot \vec{B}');$$

$$\operatorname{div}\varphi(r)\vec{A}(r) \cdot \vec{r} = \frac{\varphi'}{r}(\vec{r} \cdot \vec{A}) + \frac{\varphi}{r}(\vec{r} \cdot \vec{A}');$$

$$\operatorname{rot}\varphi(r)\vec{A}(r) \cdot \vec{r} = \frac{\varphi'}{r}[\vec{r} \times \vec{A}] + \frac{\varphi}{r}[\vec{r} \times \vec{A}'];$$

$$(\vec{p} \cdot \nabla)\varphi(r)\vec{A}(r) = \frac{\vec{p} \cdot \vec{r}}{r}(\varphi'\vec{A} + \varphi\vec{A}').$$

1.23. Prove the identities:

    (a) $\operatorname{grad}(\varphi\psi) = \varphi\operatorname{grad}\psi + \psi\operatorname{grad}\varphi$;

    (b) $\operatorname{div}(\varphi\vec{A}) = \varphi\operatorname{div}\vec{A} + \vec{A} \cdot \operatorname{grad}\varphi$;

    (c) $\operatorname{rot}(\varphi\vec{A}) = \varphi\operatorname{rot}\vec{A} - \vec{A} \times \operatorname{grad}\varphi$;

    (d) $\operatorname{div}(\vec{A} \times \vec{B}) = \vec{B} \cdot \operatorname{rot}\vec{A} - \vec{A} \cdot \operatorname{rot}\vec{B}$;

    (e) $\operatorname{rot}(\vec{A} \times \vec{B}) = \vec{A} \cdot \operatorname{div}\vec{B} - \vec{B} \cdot \operatorname{div}\vec{A} + (\vec{B} \cdot \nabla)\vec{A} - (\vec{A} \cdot \nabla)\vec{B}$;

    (f) $\operatorname{grad}(\vec{A} \cdot \vec{B}) = \vec{A} \times \operatorname{rot}\vec{B} + \vec{B} \times \operatorname{rot}\vec{A} + (\vec{B} \cdot \nabla)\vec{A} + (\vec{A} \cdot \nabla)\vec{B}$.

The scalars φ and ψ and the vectors $\vec{A}$ and $\vec{B}$ are functions of the radius-vector $\vec{r}$.

1.24. Prove the identities:

    (a) $\vec{C} \cdot \operatorname{grad}(\vec{A} \cdot \vec{B}) = \vec{A} \cdot (\vec{C} \cdot \nabla)\vec{B} + \vec{B}(\vec{C} \cdot \nabla)\vec{A}$;

    (b) $(\vec{C} \cdot \nabla)[\vec{A} \times \vec{B}] = \vec{A} \times (\vec{C} \cdot \nabla)\vec{B} - \vec{B} \times (\vec{C} \cdot \nabla)\vec{A}$;

    (c) $(\nabla \cdot \vec{A})\vec{B} = (\vec{A} \cdot \nabla)\vec{B} + \vec{B}\operatorname{div}\vec{A}$;

    (d) $[\vec{A} \times \vec{B}] \cdot \operatorname{rot}\vec{C} = \vec{B} \cdot (\vec{A} \cdot \nabla)\vec{C} - \vec{A} \cdot (\vec{B} \cdot \nabla)\vec{C}$;

    (e) $[\vec{A} \times \nabla] \times \vec{B} = (\vec{A} \cdot \nabla)\vec{B} + \vec{A} \times \operatorname{rot}\vec{B} - \vec{A}\operatorname{div}\vec{B}$;

    (f) $[\nabla \times \vec{A}] \times \vec{B} = \vec{A}\operatorname{div}\vec{B} - (\vec{A} \cdot \nabla)\vec{B} - \vec{A} \times \operatorname{rot}\vec{B} - \vec{B} \times \operatorname{rot}\vec{A}$.

1.25. Transform the volume integral

$$\int (\text{grad}\varphi \cdot \text{rot}\vec{A})dV$$

into a surface integral.
Answer.

$$\int (\text{grad}\varphi \cdot \text{rot}\vec{A})dV = \oint [\vec{A} \times \text{grad}\varphi]dS = \int \varphi \text{rot}\vec{A}dS.$$

1.26. Using the Ostrogradskii–Gauss theorem, calculate the integrals

$$\vec{I} = \oint \vec{r}(\vec{a} \cdot \vec{n})dS, \quad \vec{I} = \oint (\vec{a} \cdot \vec{r})\vec{n}dS,$$

if the volume enveloped by a closed surface is equal to V; $\vec{a}$ is a constant vector; and $\vec{n}$ is a unit vector normal to the surface.
Answer. Here, like in a number of other cases, it is convenient to consider the scalar product of the integral and a random constant vector $\vec{c}$:

$$\vec{c} \cdot \oint \vec{r}(\vec{a} \cdot \vec{n})dS = \oint (\vec{c} \cdot \vec{r})\vec{a}_n dS = \int \text{div}[(\vec{c} \cdot \vec{r}) \times \vec{a}]dV$$

$$= (\vec{a} \cdot \vec{c}) \int dV = (\vec{a} \cdot \vec{c})V.$$

Since $\vec{c}$ is a random vector, this implies that $\int (\vec{a} \cdot \vec{n})\vec{r}dS = \vec{a}V$. In the same way, we get $\oint (\vec{a} \cdot \vec{r})\vec{n}dS = \vec{a}V$.

1.27. Transform the integrals over a closed surface ($\vec{b}$ is a constant vector, and $\vec{n}$ is a unit vector normal to the surface)

$$\vec{I} = \oint \vec{n}\varphi dS, \quad \vec{I} = \oint [\vec{n} \times \vec{A}]dS, \quad \vec{I} = \oint (\vec{n} \cdot \vec{b})\vec{A}dS$$

into the integrals over the volume inside the surface.
Answer.

$$\oint \vec{n}\varphi dS = \int \text{grad}\varphi dV, \quad \oint [\vec{n} \times \vec{A}]dS = \int \text{rot}\vec{A}dV,$$

$$\oint (\vec{n} \cdot \vec{b})\vec{A}dS = \int (\vec{b} \cdot \nabla)\vec{A}dV.$$

1.28. Using the Stokes theorem, calculate the integral $\vec{I} = \oint \varphi d\vec{l}$.

Answer.

$$\oint \varphi d\vec{l} = \int [\vec{n} \times \text{grad}\varphi] dS.$$

1.29. Prove the identity

$$\int (\vec{A} \cdot \text{rot rot} \vec{B} - \vec{B} \cdot \text{rot rot} \vec{A}) dV$$

$$= \oint [[\vec{B} \times \text{rot} \vec{A}] - [\vec{A} \times \text{rot} \vec{B}]] \cdot d\vec{S}.$$

1.30. Show that $\int_V \vec{A} dV = 0$ and $A_n = 0$ if, respectively, $\text{div} \vec{A} = 0$ inside the volume V and on its boundary.

1.31. Show that the divergence of the vector

$$\vec{A} + \frac{1}{4\pi} \text{grad} \int \frac{\text{div} \vec{A}(\vec{r}')}{|\vec{r} - \vec{r}'|} dV'$$

is equal to zero.

1.32. For a three-dimensional second-rank tensor, prove the Ostrogradskii–Gauss theorem:

$$\int \frac{\partial T_{\alpha\beta}}{\partial x_\alpha} dV = \oint T_{\alpha\beta} dS_\alpha.$$

Hint. Start from the Ostrogradskii–Gauss theorem for the vector $A_\alpha = T_{\alpha\beta} \bar{a}_\beta$, where $\vec{a}$ is any constant vector.

1.33. To find a solution of the Laplace equation in a spherical coordinate system for a scalar function that depends on only one coordinate: (a) on r; (b) on ϑ; (c) on α.

Answer.

$$\text{(a)} \ A + \frac{B}{r}; \quad \text{(b)} \ A + B \ln \tan \frac{\vartheta}{2}; \quad \text{(c)} \ A + B\alpha.$$

1.34. Find a solution of the Laplace equation in a cylindrical coordinate system for a scalar function that depends on only one coordinate: (a) on r; (b) on α; (c) on z.

Answer. (a) $A + B \ln r$; (b) $A + B\alpha$; (c) $A + Bz$.

Section 2

Elements of the Special Theory
of Relativity

2.1. Lorentz transformations

The coordinates and the time in two inertial reference systems S and S' are connected with one another by the *Lorentz transformations*:

$$x = \frac{1}{\gamma}(x' + Vt'), \quad y = y', \quad z = z', \quad t = \frac{1}{\gamma}(t' + Vx'/c^2),$$

$$(2.1)$$

where $\gamma = \sqrt{1 - V^2/c^2}$.

The corresponding coordinate axes of the systems S and S' are parallel to each other, the relative velocity is directed along the axis Ox, and, at $t = t' = 0$, the coordinate origins of S and S' coincide.

The inverse Lorentz transformations can be obtained by changing the sign of the velocity V:

$$x' = \frac{1}{\gamma}(x - Vt), \quad y' = y, \quad z' = z, \quad t' = \frac{1}{\gamma}(t - Vx/c^2).$$

The quantities $x^0 = ct$, $x^1 = x$, $x^2 = y$, $x^3 = z$ are the coordinates of a world point

$$x^i = (ct, \vec{r}).$$

Any four quantities A^0, A^1, A^2, A^3 that transform from one inertial reference system to another in the same way as the coordinates and the time, i.e., by the formulas

$$A^0 = \frac{1}{\gamma}(A^{0\prime} + \beta\, A^{1\prime}), \quad A^1 = \frac{1}{\gamma}(A^{1\prime} + \beta\, A^{0\prime}), \quad A^2 = A^{2\prime}, \quad A^3 = A^{3\prime}$$

$$(2.2)$$

$(\beta = V/c)$, form *a four-dimensional vector* A^i, $i = 0, 1, 2, 3$.

The three-dimensional vector $\vec{A} = (A^1,\ A^2,\ A^3)$ and the quantity A^0 are called, respectively, the *spatial* and *temporal components* of the four-dimensional vector A^i.

For the sake of convenience, let us introduce two types of components of four-dimensional vectors, which are denoted as follows: A^i and A_i. In this case,

$$A_0 = A^0, \quad A_1 = -A^1, \quad A_2 = -A^2, \quad A_3 = -A^3.$$

The quantities A^i and A_i are called, respectively, *contravariant* and *covariants* components of a four-dimensional vector. The square of a four-dimensional vector is written as

$$(A_i)^2 = (A_0)^2 - (A_1)^2 - (A_2)^2 - (A_3)^2$$

or

$$\sum_{i=0}^{3} A^i A_i = A^0 A_0 + A^1 A_1 + A^2 A_2 + A^3 A_3.$$

Similar sums are denoted as $A^i A_i$, by omitting the sign of sum.

The scalar product of two four-dimensional vectors is defined as follows:

$$A^i B_i = A^0 B_0 + A^1 B_1 + A^2 B_2 + A^3 B_3. \tag{2.3}$$

The squares of four-dimensional vectors $A^i A_i$ and their scalar products $A^i B_i$ have identical values in all inertial reference systems (invariants under the Lorentz transformations). The four-dimensional vector A_i is called *space-like* if $A^i A_i < 0$, or *time-like* if $A^i A_i > 0$.

The invariant quantity

$$s_{12} = [c^2(t_2 - t_1)^2 - (x_2 - x_1)^2 - (y_2 - y_1)^2 - (z_2 - z_1)^2]^{1/2}$$

is called the interval between two events with the coordinates (x_1, y_1, z_1, t) and (x_2, y_2, z_2, t).

The time counted by a clock that moves together with the object is called the *intrinsic time* of this object.

If the object moves relative to the system S with a velocity v, then the intrinsic time interval $d\tau$ can be written in terms of the time interval dt in the system S by the formula

$$d\tau = dt\sqrt{1 - v^2/c^2}.$$

The quantity $dt\sqrt{1 - \beta^2}$ is an invariant under the Lorentz transformations.

If a rod has a length l_0 in its rest system, then, during the motion with a velocity v along its axis, this rod has the length

$$l = l_0\sqrt{1 - v^2/c^2}$$

in the immovable reference system.

The *four-dimensional velocity* of a particle is called a four-dimensional vector, whose components are defined by the formula

$$u^i = \frac{dx^i}{ds} = \left(\frac{1}{\sqrt{1 - v^2/c^2}}, \frac{\vec{v}}{c\sqrt{1 - v^2/c^2}} \right), \qquad (2.4)$$

where $\vec{v} = d\vec{r}/dt$ is the ordinary velocity of a particle.

Formula (2.4) yields

$$u^i u_i = 1.$$

The four-dimensional velocity, as any other four-dimensional vector, is transformed by formulas (2.2).

Components of the ordinary velocity are not spatial components of any four-dimensional vector and are transformed by the formulas $(\vec{V} \| x)$:

$$v_x = \frac{v'_x + V}{1 + v'_x V/c^2}, \quad v_y = \frac{v'_y \sqrt{1 - V^2/c^2}}{1 + v'_x V/c^2}, \quad v_z = \frac{v'_z \sqrt{1 - V^2/c^2}}{1 + v'_x V/c^2}.$$

If the velocity of a particle is directed to the axis x at the angles ϑ and ϑ' in the systems S and S', respectively, then

$$\tan\vartheta = \frac{v'\sqrt{1 - V^2/c^2}\sin\vartheta'}{v'\cos\vartheta' + V}; \quad v' = \sqrt{v_x'^2 + v_y'^2 + v_z'^2}.$$

The *four-dimensional acceleration* of a particle is called a four-dimensional vector with the components

$$w^i = \frac{du^i}{ds} = \frac{d^2x^i}{ds^2}.$$

The wave vector $\vec{k}$ and the frequency ω of a plane electromagnetic wave are components of the four-dimensional *wave vector* k^i:

$$k^i = (\omega/c,\ \vec{k}).$$

Therefore, the phase of a plane wave $\varphi = -k^i x_i = \vec{k}\vec{r} - \omega t$ is an invariant.

Formulas (2.2) yield the formula of transformation of the angle ϑ between a light ray and the axis x:

$$\tan\vartheta = \frac{\sin\vartheta'}{\gamma(\cos\vartheta' + \beta)} \quad \text{or} \quad \cos\vartheta = \frac{\cos\vartheta' + \beta}{1 + \beta\cos\vartheta'}.$$

* * *

Example 2.1. Let the periodic process of reflection of a light ray from two mirrors fixed on the ends of a rod with length l be used for the time measurement. One period includes the duration of the motion of a ray from one mirror to another one and in the reverse direction. The light-based clock is immovable in the system S' and is oriented in parallel to the direction of motion. Using the postulate of invariability of the velocity of light, show that the intrinsic time interval $d\tau$ is presented in terms of a time interval dt in the system S by the formula $d\tau = dt\sqrt{1 - V^2/c^2}$.

Solution. In the system S', the period $T' = 2l/c$. In the system S, the time T_1 of the motion of a light ray along the rod in the direction of the relative velocity V is calculated from the equation

$$T_1 = \frac{1}{c}\left(l\sqrt{1 - V^2/c^2} + VT_1\right),$$

The time of the motion in the reverse direction T_2 can be determined, by replacing V in the last formula by $-V$. Now, the ratio of T' to $T = T_1 + T_2$ is

$$\frac{T'}{T} = \sqrt{1 - V^2/c^2},$$

which yields

$$d\tau = dt\sqrt{1 - V^2/c^2}.$$

Example 2.2. Obtain the formulas of the Lorentz transformation from the system S' to the system S for the radius-vector $\vec{r}$ and the time t, not considering that the velocity $\vec{V}$ of the system S' relative to S is parallel to the axis x. Present the result in the vector form.

Hint. Decompose $\vec{r}$ into the longitudinal and transverse components relative to $\vec{V}$ and use the Lorentz transformations (2.1).

Solution. Let us decompose $\vec{r}$ into the tangent (the direction of $\vec{r}_{\parallel}$ at any time moment coincides with the direction of the velocity $\vec{V}$) and normal ($\vec{r}_{\perp}$ is perpendicular to $\vec{V}$) components:

$$\vec{r} = \vec{r}_{\parallel} + \vec{r}_{\perp}.$$

Analogously, for $\vec{r}'$ (Fig. 2.1), we have

$$\vec{r}' = \vec{r}'_{\parallel} + \vec{r}'_{\perp}.$$

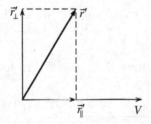

Fig. 2.1

By the rules of vector analysis, we now form the vectors in the required directions:

$$\vec{r}_\parallel' = \vec{V}\frac{\vec{r}' \cdot \vec{V}}{V^2}; \quad \vec{r}_\perp' = \lambda\frac{[\vec{V} \times \vec{r}'] \times \vec{V}}{V^2};$$

here, $\vec{n} = \vec{V}/V$ is a unit vector, whose direction coincides with that of the vector $\vec{V}$; and λ is the coefficient of proportionality, which is chosen so that the length of the vector is proper.

Let us find λ. For this purpose, we rewrite the perpendicular component of the radius-vector $\vec{r}'$,

$$\vec{r}_\perp' = \lambda[[\vec{n} \times \vec{r}'] \times \vec{n}],$$

and find its length:

$$|\vec{r}_\perp'| = \lambda|[[\vec{n} \times \vec{r}'] \times \vec{n}]| = \lambda \cdot |[\vec{n} \times \vec{r}']| \cdot |\vec{n}| \cdot \sin\frac{\pi}{2} = \lambda \cdot |[\vec{n} \times \vec{r}']|$$

$$= \lambda \cdot |\vec{n}| \cdot |\vec{r}'| \cdot \sin\vartheta = \lambda \cdot |\vec{r}'| \cdot \sin\vartheta.$$

On the other hand, it is seen from Fig. 2.1 that

$$|\vec{r}_\perp'| = |\vec{r}'| \cdot \sin\vartheta.$$

Hence, $\lambda = 1$. Thus,

$$\vec{r}_\perp' = \frac{[\vec{V} \times \vec{r}] \times \vec{V}}{V^2}.$$

We now apply the Lorentz transformation to $\vec{r}_\parallel$ and $\vec{r}_\perp$:

$$\vec{r}_\parallel = \frac{\vec{r}_\parallel' + \vec{V}t'}{\sqrt{1 - \frac{V^2}{c^2}}}; \quad \vec{r}_\perp = \vec{r}_\perp'. \tag{2.5}$$

Substituting the values of $\vec{r}_\parallel'$ and $\vec{r}_\perp'$ in formulas (2.5), we get

$$\vec{r} = \frac{\vec{V}\frac{\vec{r}' \cdot \vec{V}}{V^2} + \vec{V}t}{\sqrt{1 - \frac{V^2}{c^2}}} + \frac{[\vec{V} \times \vec{r}] \times \vec{V}}{V^2}.$$

The Lorentz transformation of the time takes the form

$$t = \frac{t' + \frac{\vec{r}' \cdot \vec{V}}{c^2}}{\sqrt{1 - \frac{V^2}{c^2}}}.$$

Example 2.3. A beam of light in some reference system forms a solid angle $d\Omega$. How will this angle be changed in another inertial reference system?

Solution. Consider the beam of light in the system S inside of the solid angle $d\Omega$. In the spherical coordinate system, the solid angle $d\Omega = \sin \vartheta d\vartheta d\alpha$.

In the system S', this beam is observed inside of the angle $d\Omega' = \sin \vartheta' d\vartheta' d\alpha' = -d(\cos \vartheta') d\alpha'$.

The angle $\alpha = \alpha'$. Respectively, $d\alpha = d\alpha'$.

$$\cos \vartheta' = \frac{\cos \vartheta - \frac{V}{c}}{1 - \frac{V}{c} \cos \vartheta};$$

$$d(\cos \vartheta') = \frac{d(\cos \vartheta)\left(1 - \frac{V}{c}\cos\vartheta\right) + \left(\cos\vartheta - \frac{V}{c}\right)d(\cos\vartheta)}{\left(1 - \frac{V}{c}\cos\vartheta\right)^2}$$

$$= d(\cos\vartheta)\frac{\left(1 - \frac{V}{c}\cos\vartheta + \frac{V}{c}\cos\vartheta - \frac{V^2}{c^2}\right)}{\left(1 - \frac{V}{c}\cos\vartheta\right)^2}$$

$$= -\sin\vartheta \frac{1 - \frac{V^2}{c^2}}{\left(1 - \frac{V}{c}\cos\vartheta\right)^2};$$

$$d\Omega' = \frac{\sin\vartheta\left(1 - \frac{V^2}{c^2}\right)}{\left(1 - \frac{V}{c}\cos\vartheta\right)^2}d\vartheta d\alpha = \frac{1 - \frac{V^2}{c^2}}{\left(1 - \frac{V}{c}\cos\vartheta\right)^2}d\Omega.$$

Problems

2.1. The length of a rod, which moves along its axis in some reference system, can be determined in the following way: measure the time interval, for which the rod passes through a fixed point of this system, and multiply it by the velocity of the rod. Show that the ordinary Lorentz reduction of a length is observed for such method of measurement.

2.2. The system S' moves relative to the system S with velocity V. At the moment in time when the coordinate origins coincide, the clocks in each coordinate system point out the same time: $t = t' = 0$. Which coordinates will the world point have in each of these systems that possesses the following property: the clocks located at this point in the systems S and S' point out the same time $t = t'$? Determine the law of motion of this point.

Answer. The coordinates of the clocks pointing out the same time $t = t'$ are, in the systems S and S',

$$ x = \frac{c^2}{V}\left(1 - \frac{1}{V}\right) t, \quad x' = -\frac{c^2}{V}\left(1 - \frac{1}{V}\right) t. \quad (2.6) $$

It is seen from formulas (2.6) that the point at which $t = t'$ moves uniformly in each of the systems S and S'. If we introduce the reference system, relative to which this point is immovable, then S and S' move in the opposite sides with the same velocity $V_0 = \frac{c^2}{V}(1 - \frac{1}{V})$.

2.3. Let the periodic process of reflection of a light ray from two mirrors fixed on the ends of a rod with length l be used to measure the time. One period is the time of motion of a ray from one mirror to the other and in the reverse direction. Using the postulate of invariability of the velocity of light, show that the intrinsic time interval $d\tau$ is presented in terms of a time interval dt in the system S by the formula $d\tau = dt\sqrt{1 - V^2/c^2}$. The light-based clock is immovable in the system S' and is oriented: a) in parallel to the direction of motion; b) perpendicularly to the direction of motion of the rod.

Answer. In the system S', the duration of one period $T' = 2l/c$; in the system S, the time of motion T_1 of a "light spot" along the rod in the direction along the velocity V, is calculated by the equation

$$ T_1 = \frac{1}{c}(l\sqrt{1 - V^2/c^2} + VT_1). $$

The time of motion in the reverse direction T_2 is determined by the replacement of V by $-V$.

For the ratio of T' to $T = T_1 + T_2$, we get

$$\frac{T'}{T} = \sqrt{1 - V^2/c^2},$$

which yields

$$d\tau = dt\sqrt{1 - V^2/c^2}.$$

2.4. The train $A'B'$, whose length $l_0 = 8.64 \cdot 10^8$ km in its rest system, moves with the velocity $V = 240000$ km/s along a platform with the same length in its rest system. On the head B' and the end A' of the train, two identical clocks synchronized with each other are placed. The same clocks are mounted on the ends A and B of the platform. At the moment in time when the train head passes near the beginning A of the platform, the corresponding clocks show 12 h 00 min. Answer the questions: (a) Can we assert that, at this moment in time in some reference system, all clocks show 12 h 00 min?; (b) What time does each of the clocks show at the moment in time when the train end passes near the beginning A of the platform?; (c) What time does each of the clocks show at the moment in time when the train head passes near the end B of the platform?

Answer. (a) We cannot. The time of 12 h 00 min can be shown simultaneously by two clocks in one of the reference systems and only one clock in another reference system.

(b) The readings of the clocks, which coincide in the space, are independent of the choice of a reference system:

$$t_{A'} = 12\,\text{h}\,00\,\text{min} + \frac{l_0}{V} = 13\,\text{h}\,00\,\text{min};$$

$$t_A = 12\,\text{h}\,00\,\text{min} + \frac{l_0}{V}\sqrt{1 - \frac{V^2}{c^2}} = 12\,\text{h}\,36\,\text{min}.$$

The readings of the remaining clocks B and B' depend on the choice of a reference system due to the relativity of simultaneity.

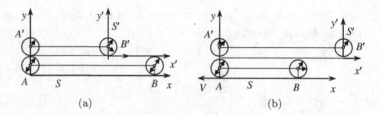

Fig. 2.2

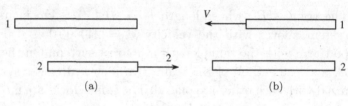

Fig. 2.3

From the observation point on the platform (Fig. 2.2, (a)):

$$t_{B'} = 12 \text{ h } 21.6 \text{ min}, \quad t_B = t_A = 12 \text{ h } 36 \text{ min.}$$

From the observation point in the train (Fig. 2.2, (b)):

$$t_{B'} = t_{A'} = 13 \text{ h } 00 \text{ min}, \quad t_B = 13 \text{ h } 14.4 \text{ min.}$$

(c) From the observation point on the platform:

$$t_A = 13 \text{ h } 00 \text{ min} = t_B, \quad t_{B'} = 12 \text{ h } 36 \text{ min},$$
$$t_{A'} = 13 \text{ h } 14.4 \text{ min.}$$

From the observation point in the train:

$$t_A = 12 \text{ h } 21.6 \text{ min}, \quad t_{A'} = t_{B'} = 12 \text{ h } 36 \text{ min},$$
$$t_B = 13 \text{ h } 00 \text{ min.}$$

In all cases, those clocks lag, whose readings should be compared with those of two clocks in another reference system.

2.5. Let two identical length standards (Fig. 2.3, 1 and 2) with length l_0 in their rest system move uniformly in opposite directions in parallel to the common axis x. The observer connected with one of the length standards noticed that the time interval between the coincidence of left and right ends is equal to Δt. What is the relative velocity v of the length standards? In which sequence do their ends coincide for the observers connected with each of the length standards, and for the observer relative to which both objects move with the same velocity in the reverse direction?

Answer. $v = \frac{2l_0 \Delta t}{(\Delta t)^2 + l_0^2/c^2}$.

For the observer connected with length standard *1* (Fig. 2.3, (a)), the left ends coincide first, and then the right ends coincide. For the observer connected with length standard *2* (Fig. 2.3, (b)), the reverse happens. From the observation point, relative to which the length standards move with the same velocity, the ends coincide simultaneously.

2.6. Write the formulas of the Lorentz transformation for any four-dimensional vector $A_i = (A_0, \vec{A})$, without assuming that the velocity of the system S' relative to S is parallel to the axis x.

Answer.

$$A = \gamma^{-1} \left(\vec{A}' + \frac{\vec{V}}{c} A_0' \right) + (\gamma^{-1} - 1) \frac{[\vec{A}' \times \vec{V}] \times \vec{V}}{V^2},$$

$$A_0 = \gamma^{-1} \left(A_0' + \frac{\vec{A}' \cdot \vec{V}}{c} \right).$$

2.7. Deduce the formulas of summation of velocities in the case where the velocity $\vec{V}$ of the system S' relative to S has an arbitrary direction. Write the formulas in vector form.

Answer.

$$\vec{v} = \vec{v}_\parallel + \vec{v}_\perp = \frac{\vec{v}' + \vec{V} + (\gamma - 1) \frac{\vec{V}}{V^2} [(\vec{v}' \cdot \vec{V}) + V^2]}{\gamma \left(1 + \frac{\vec{v}' \cdot \vec{V}}{c^2} \right)},$$

where $\vec{v}$ and $\vec{v}'$ are the velocities in the systems S and S'.

2.8. Let us set three reference systems: S, S', S''. The system S'' moves relative to S' in parallel to the axis x' with velocity V', S' moves relative to S in parallel to the axis x with velocity V.

The corresponding axes of all three systems are parallel. Write the Lorentz transformation from S'' to S and obtain the formula of summation of parallel velocities.

2.9. Prove the formula

$$\sqrt{1 - \frac{v^2}{c^2}} = \frac{\sqrt{1 - v'^2/c^2} - \sqrt{1 - V^2/c^2}}{1 + \vec{v}' \cdot \vec{V}/c^2},$$

where $\vec{v}$ and $\vec{v}'$ are the velocities of a particle in the systems S and S'; and $\vec{V}$ is the velocity of S' relative to S.

2.10. Prove the relation

$$v = \frac{\sqrt{(\vec{v}' + \vec{V})^2 - [\vec{v}' \times \vec{V}]^2/c^2}}{1 + \vec{v}' \cdot \vec{V}/c^2},$$

where $\vec{v}$ and $\vec{v}'$ are the velocities of a particle in the systems S and S'; and $\vec{V}$ is the velocity of S' relative to S.

2.11. Two length standards with the same length l_0 in their rest system approach each other, by moving in opposite directions with the same velocities v relative to some reference system. What is the length l of each of the length standards, and what is measured in the reference system connected with another length standard?

Answer.

$$l = l_0 \frac{1 - v^2/c^2}{1 + v^2/c^2}.$$

2.12. Two beams of electrons approach each other, moving in opposite directions with velocities $v = 0.9\ c$ relative to the laboratory coordinate system. What is the relative velocity V of electrons: (a) from the observation point in the laboratory coordinate system; (b) from the observation point which moves together with one of the beams of electrons?

Answer. (a) $V = 2 \cdot 0.9\ c = 1.8\ c$; (b) $V = 0.994\ c$.

2.13. Deduce the transformation formulas for the acceleration $\dot{\vec{v}}$ in the case where the system S' moves relative to the system S with an arbitrarily directed velocity $\vec{V}$. Present these formulas in vector form.

Answer.

$$\dot{\vec{v}} = \frac{1}{\gamma^2 s^2}\dot{\vec{v}}' - \frac{(\gamma-1)(\dot{\vec{v}}' \cdot \vec{V})\vec{V}}{\gamma^3 s^3 V^2} - \frac{(\dot{\vec{v}}' \cdot \vec{V})\vec{v}'}{\gamma^2 s^3 c^2},$$

where $s = 1 + \frac{\vec{v}' \cdot \vec{V}}{c^2}$.

2.14. Present the components of a four-dimensional acceleration w^i in terms of the ordinary acceleration $\dot{\vec{v}}$ and the velocity $\vec{v}$ of a particle. Find w^i. Will the four-dimensional acceleration be time-like or space-like?

Answer.

$$w^i w_i = -\gamma^6 \left(\dot{\vec{v}}^2 - \left[\dot{\vec{v}} \times \frac{\vec{v}}{c}\right]^2 \right) = -\gamma^4 \left(\dot{\vec{v}}^2 + \gamma^2 \dot{\vec{v}}^2 \frac{\vec{v}^2}{c^2} \right) < 0,$$

i.e., the four-dimensional acceleration is a space-like vector.

2.15. Present the acceleration $\dot{\vec{v}}'$ of a particle in the inertial system instantly associated with it in terms of its acceleration $\dot{\vec{v}}$ in the laboratory system. Consider the cases where the velocity $\vec{v}$ of a particle varies only by magnitude or only by direction.

Answer. If the velocity of a particle varies only by magnitude, then $\dot{\vec{v}}\|\vec{v}$ and $\dot{\vec{v}}' = \gamma^3 \dot{\vec{v}}$.

If the velocity of a particle varies only by direction, then $\vec{v} \perp \dot{\vec{v}}$ and $\vec{v} \cdot \dot{\vec{v}} = 0$, hence, $\dot{\vec{v}}' = \gamma^2 \dot{\vec{v}}$.

2.16. A relativistic particle undergoes uniformly accelerated one-dimensional motion (acceleration $\dot{v} \equiv w$ is invariable in the own reference system). Find the dependence of the velocity $v(t)$ and the coordinates $x(t)$ of the particle on the time t in the laboratory reference system, if the initial velocity is v_0, and if the initial coordinate is x_0. Consider, in particular, the nonrelativistic and ultrarelativistic cases.

Answer.

$$v(t) = \frac{wt + v_0/\sqrt{1 - (v_0/c)^2}}{\sqrt{1 + |(wt)^2/c^2 + (v_0/c)^2/(1 - v_0^2/c^2)|^2}};$$

$$x(t) = \frac{c^2}{\omega}\left\{ \sqrt{1 + \frac{1}{c^2}\left(wt + \frac{v_0}{1 - v_0^2/c^2}\right)^2} - \frac{v_0/c}{1 - v_0^2/c^2} \right\} + x_0.$$

In the ultrarelativistic case:

$$v(t) \approx c, x(t) \approx ct - c^2/\omega.$$

In the nonrelativistic case:

$$v(t) = v_0 + \omega t, x(t) = x_0 + v_0 t + 1/2\omega t^2.$$

2.17. A rocket is accelerated from rest to the velocity $v = \sqrt{0.9999}\,c$. The acceleration of the rocket is $|\dot{\vec{v}}| = 20$ m/s^2 in the system that instantly accompanies the rocket.

How much time does the acceleration of the rocket continue by the clocks in the immovable reference system and in the rocket?

Hint. Neglect the action of inertial forces on the course of the clock in the rocket.

Answer. The acceleration duration by the clock in the immovable system:

$$T = \frac{1}{|\dot{\vec{v}}|} \int_0^V \frac{dv}{(1 - v^2/c^2)^{3/2}} = \frac{v}{|\dot{\vec{v}}|\sqrt{1 - v^2/c^2}} = 47.5 \text{ yr}.$$

The acceleration duration by the clock in the system connected with the rocket,

$$\tau = \frac{c}{2|\dot{\vec{v}}|} \ln \left| \frac{1 + v/c}{1 - v/c} \right| = 2.5 \text{ yr}.$$

2.18. Let the system S' with velocity $\vec{V}$ and two bodies with velocities $\vec{v}_1$ and $\vec{v}_2$ move relative to the system S. What angle α between the velocities of these bodies in the system S and the system S' will be observed? What behavior of the angle α between the velocities of two bodies will be observed in the case where the velocity of the system S' relative to S approaches c?

Answer. In the system S: $\cos\alpha = \frac{\vec{v}_1 \cdot \vec{v}_2}{|\vec{v}_1||\vec{v}_2|}$. In the system S':

$$\cos\alpha' = \frac{(\vec{v}_1 - \vec{V}) \cdot (\vec{v}_2 - \vec{V}) - \frac{1}{c^2}[\vec{v}_1 \times \vec{V}] \cdot [\vec{v}_2 \times \vec{V}]}{\sqrt{(\vec{v}_1 - \vec{V})^2 - \frac{1}{c^2}[\vec{v}_1 \times \vec{V}]^2}\sqrt{(\vec{v}_2 - \vec{V})^2 - \frac{1}{c^2}[\vec{v}_2 \times \vec{V}]^2}}.$$

2.19. At some moment in time, the direction of a light ray from a star forms the angle θ with the orbital velocity $\vec{v}$ of the Earth (in the coordinate system connected with the Sun). Find the change of the direction from the Earth to the star for a half-year (aberration of light), not considering the approximations related to the smallness of v/c.

2.20. Let us assume that the stars in the part of the Galaxy nearest to us are uniformly distributed. Which distribution $d/d\Omega'$ will be observed by an observer in a rocket moving with a velocity close to the velocity of light?

Answer. $\frac{dN}{d\Omega'} = \frac{N_0}{4\pi} \cdot \frac{d\Omega}{d\Omega'} = \frac{N_0}{4\pi} \frac{1-\beta^2}{(1-\beta\cos\vartheta')^2}$, where N_0 is the total number of visible stars.

2.21. Find the transformation formulas of the frequency ω (Doppler effect) and the wave vector $\vec{k}$ of a plane monochromatic wave of light when transiting from one inertial system to another one. The direction of the relative velocity $\vec{V}$ is arbitrary.

Answer.

$$\omega = \gamma\omega'\left(1 + \frac{\vec{n}' \cdot \vec{V}}{c}\right) \quad \text{or} \quad \omega = \frac{\omega'}{\gamma\left(1 - \frac{\vec{n}\cdot\vec{V}}{c}\right)},$$

$$\vec{k} = \gamma\left(\vec{k'} + \frac{\vec{V}\omega'}{c^2}\right) + (\gamma - 1)[\vec{k'} \times \vec{V}] \times \frac{\vec{V}}{V^2},$$

where $\vec{n} = \vec{k}/k$, $\vec{n}' = \vec{k'}/k$, $k = \omega/c$.

2.22. Find the frequency of a wave of light, which is observed by the transverse Doppler effect (the light propagation direction is perpendicular to the direction of motion of a light source in the system associated with a light receiver). Which propagation direction of the wave is observed in the system associated with the source?

Answer. If ω_0 is the frequency in the system, where the source is at rest, and if V is the velocity of the source relative to the light receiver, then the receiver sees a lower frequency $\omega = \omega_0\sqrt{1 - V^2/c^2}$ (red shift). The angle α of a ray with the direction of motion of the source in its rest system is given by the formula

$$\cos \alpha = -\frac{V}{c}.$$

The angle α is close to $90°$ only if $V \ll c$. If $V \to c$, then $\alpha \to \pi$.

2.23. The wavelength of light emitted by some source in the system, where the source is at rest, is equal to λ_0. What wavelength λ will be registered by: (a) an observer approaching the source with velocity V; (b) an observer moving away from the source with the same velocity?

Answer.

$$\text{(a) } \lambda = \lambda_0 \sqrt{\frac{1 - V/c}{1 + V/c}}; \quad \text{(b) } \lambda = \lambda_0 \sqrt{\frac{1 + V/c}{1 - V/c}}.$$

2.24. A mirror moves: (a) perpendicularly to its own plane with velocity $\vec{V}$; (b) laterally along its own plane with velocity $\vec{V}$. Find the law of reflection of a plane monochromatic wave from such mirror, which replaces the law of equality of the angles of incidence and reflection for $V = 0$, and the law of transformation of the frequency of the wave at its reflection. Consider, in particular, the case where $V \to c$.

2.25. Introduce a four-dimensional wave vector describing the propagation of a plane monochromatic wave in a medium with a refractive index n, which moves with velocity V (phase velocity of the wave in an immovable medium $v' = c/n$). Find the formulas of transformation of the frequency and the angle between the wave vector and the motion direction of the medium and the phase velocity of the wave.

2.2. Four-dimensional vectors and tensors

When transiting from an inertial system to another one, the components of a four-dimensional vector are transformed by the formula

$$A^i = \alpha_k^i \cdot A'^k, \tag{2.7}$$

where the matrix of transformation α_k^i takes the form

$$\alpha_k^i = \begin{pmatrix} \gamma & -\beta\gamma & 0 & 0 \\ \beta\gamma & -\gamma & 0 & 0 \\ 0 & 0 & -1 & 0 \\ 0 & 0 & 0 & -1 \end{pmatrix}.$$

It corresponds to transformation (2.1), where the similar coordinate axes of the systems S and S' are parallel, the relative velocity is directed along the axis x, and the coordinate origins of both systems coincide at $t = t' = 0$.

The matrix of transformation satisfies the relations

$$\alpha^{il}\alpha_l^k = g^{ik}, \quad \alpha_l^i\alpha^{lk} = g^{ik},$$

where g^{ik} is the metric tensor, which takes the form

$$g^{ik} = \begin{pmatrix} 1 & 0 & 0 & 0 \\ 0 & -1 & 0 & 0 \\ 0 & 0 & -1 & 0 \\ 0 & 0 & 0 & -1 \end{pmatrix}. \tag{2.8}$$

The signs on the principal diagonal of the metric tensor correspond to those in formula (2.3), which defines the scalar product of two four-dimensional vectors.

The transformation inverse to formula (2.7) is written in the following way:

$$A'^i = \alpha_i^k A_k. \tag{2.9}$$

The coordinates of the world point form a four-dimensional vector and are transformed by formulas (2.7), (2.9).

For the successive implementation of two Lorentz transformations, the corresponding matrices are multiplied by the customary rule of multiplication of matrices.

A four-dimensional second-rank tensor is called the collection of 16 quantities A^{ik}, which are transformed under a transformation of coordinates as the products of components of two four-dimensional vectors. The four-dimensional tensors of higher ranks are defined analogously.

Components of a four-dimensional second-rank tensor can be represented as contravariant A^{ik}, covariant A_{ik}, and mixed A^i_k or A^k_i (they should be distinguished by observing which index (first or second) is upper or lower). The connection between different types of components is determined by the general rule: the raising or lowering of the time index (0) does not change the sign of a component, whereas the raising or lowering of the space index (1, 2, 3) changes it:

$$A_{00} = A^{00}, A_{01} = -A^{01}, A_{11} = A^{11}, \ldots,$$
$$A^0_0 = A^{00}, A^1_0 = A^{01}, A^0_1 = -A^{01}, A^1_1 = -A^{11}, \ldots$$

Tensor A^{ik} is called *symmetric*, if $A^{ik} = A^{ki}$, and *antisymmetric*, if $A^{ik} = -A^{ki}$. All diagonal components of an antisymmetric tensor are zero. The mixed components A^i_k and A^k_i of a symmetric tensor coincide. Therefore, they can be written as A^k_i, by placing the indices one over another.

It is possible to construct a scalar from components of the tensor A^{ik} by the summation:

$$A^i_i = A^0_0 + A^1_1 + A^2_2 + A^3_3$$

(in this case, of course, $A^i_i = A^i_i$). Such sum is called *the trace of a tensor*, and this operation is called a *contraction* or *simplification* of a tensor.

The operation of contraction gives the scalar product of two four-dimensional vectors, i.e., it forms the scalar $A^i B_i$ from the tensor $A^i B_k$. Generally, any contraction of a pair of indices decreases the rank of a tensor by 2.

The unit four-dimensional tensor is called the tensor δ^i_k, which satisfies the equality

$$\delta^i_k A^k = A^i$$

for any four-dimensional vector A^i. It is obvious that the components of this tensor are

$$\delta^i_k = \begin{cases} 1, & \text{if } i = k, \\ 0, & \text{if } i \neq k. \end{cases}$$

The trace $\delta^i_i = 4$. By raising one index or by lowering another one of the tensor δ^i_k, we obtain a contravariant or covariant tensor, which is denoted by g_{ik} or g^{ik} and is called *the metric tensor*. The tensors g_{ik} and g^{ik} have identical components, which are defined by relation (2.8). The index i enumerates rows, and the index k enumerates columns in the order 0, 1, 2, 3. It is obvious that

$$g_{ik}A^k = A_i, \quad g^{ik}A_k = A^i.$$

Then the scalar product of two four-dimensional vectors can be written in the form

$$A^i A_i = g_{ik}A^i A^k = g^{ik}A_i A_k.$$

The components of the tensors δ^i_k, g_{ik}, and g^{ik} are identical in all coordinate systems. The same property is possessed by the completely antisymmetric unit four-dimensional tensor of the fourth rank e^{iklm}. Its components change sign under the permutation of any two indices, and the nonzero components are equal to ± 1. The components of this tensor, for which at least two indices coincide, are zero. Only those components for which all four indices are different are nonzero. We set $e^{0123} = +1$ (in this case, $e_{0123} = -1$). Then all nonzero components e^{iklm} are $+1$ or -1. The number of such components is $4! = 24$. Therefore, $e^{iklm}e_{iklm} = -24$.

Under rotations of the coordinate system, the quantities e^{iklm} behave as components of a tensor. However, if the sign of one (or three) coordinate changes, the components of e^{iklm} remain invariable, since they are defined identically for all coordinate systems. At the same time, the components of a tensor must change sign. Therefore, e^{iklm} is not a tensor, but a pseudotensor. *Pseudotensors* of any rank, in particular, *pseudoscalars*, behave as tensors under all transformations of coordinates, except for those which cannot be reduced to rotations, i.e., except for the changes of the signs of coordinates.

The completely antisymmetric unit pseudotensor is a collection of quantities $e_{\alpha\beta\gamma}$ that change sign under the permutation of any two indices. Only the components of $e_{\alpha\beta\gamma}$ with three different indices are nonzero. In this case, we set $e_{xyz} = 1$, and all other components are

+1 or −1, depending on the number (even or odd) of permutations which reduce the sequence α, β, γ to x, y, z.

The products $e_{\alpha\beta\gamma}e_{\lambda\mu\nu}$ form the true three-dimensional tensor of the sixth rank and, therefore, can be presented in the form of combinations of the products of components of the unit three-dimensional tensor $\delta_{\alpha\beta}$.

Under reflection of the coordinate system, i.e., under the change of signs of all coordinates, the components of an ordinary three-dimensional vector also change sign. Such vectors are called *polar*. But the components of some vectors, which can be presented as the vector product of two polar vectors, do not change sign under reflection. Such vectors are called *axial*. The scalar product of polar and axial vectors is not a true scalar, but a pseudoscalar: it changes sign under reflection of the coordinates.

* * *

Problems

2.26. Prove the equalities:

$$A_i = g_{ik}A^k, \quad A_iB_i = A^i g_{ik}B^k, \quad g_{ik}g^{kl} = \delta_i^l, \quad \delta_i^i = 4,$$

where g_{ik} is the metric tensor (2.8); A_i and B_i are four-dimensional vectors. In the determination of a sum over two repeated indices, the rule of signs should be used.

2.27. Show that the tensor g_{ik} has the same components in all inertial coordinate systems.

2.28. Show that the components A^1, A^2, A^3 of the four-dimensional vector $A^i = (A^0, A^1, A^2, A^3)$ under spatial rotations are transformed as the components of a three-dimensional vector $\vec{A} = (A^1, A^2, A^3)$, and the component A^0 is a three-dimensional scalar.

2.29. Find the three-dimensional tensors into which a four-dimensional second-rank tensor is split under spatial rotations.

Answer. It is split into the three-dimensional second-rank tensor $A_{\alpha\beta}(\alpha, \beta = 1, 2, 3)$, two three-dimensional vectors $A_{0\alpha}$ and $A_{\alpha 0}(\alpha = 1, 2, 3)$, and a three-dimensional scalar A_{00}.

2.30. Show that the components of an antisymmetric four-dimensional second-rank tensor are transformed under spatial rotations as the components of two independent three-dimensional vectors.

Answer. An antisymmetric four-dimensional tensor A_{ik} can be presented in the form

$$A_{ik} = \begin{pmatrix} 0 & -B_1 & -B_2 & -B_3 \\ B_1 & 0 & A_1 & -A_2 \\ B_2 & -A_3 & 0 & A_1 \\ B_3 & A_2 & -A_1 & 0 \end{pmatrix},$$

where $A = (A_1, A_2, A_3)$ and $B = (B_1, B_2, B_3)$ are three-dimensional vectors (to be more exact, B is a polar vector, and A is an axial one).

2.31. Prove that the quantity e^{iklm}, whose definition is given in this subsection, really transforms as a pseudotensor.

2.32. Prove the equalities: (a) $e^{iklm} e_{prlm} = -2(\delta^i_p \delta^k_r - \delta^i_r \delta^k_p)$, (b) $e^{iklm} e_{pklm} = -6\delta^i_p$.

2.33. Prove the equality

$$e^{iklm} e_{lmrs} A_i B_k C^r D^s = 2(A_i D^i)(B_k C^k) - 2(A_i C^i)(B_k D^k).$$

2.34. Compose a four-dimensional vector from the partial derivatives $\partial \varphi / \partial x^i$ ($i = 0, 1, 2, 3$), where φ is a scalar quantity. Find the expression for components ∇_i of the operator of four-dimensional gradient.

Answer. The invariant quantity (four-dimensional differential)

$$d\varphi = \frac{\partial \varphi}{\partial x^0} dx^0 + \frac{\partial \varphi}{\partial x^1} dx^1 + \frac{\partial \varphi}{\partial x^2} dx^2 + \frac{\partial \varphi}{\partial x^3} dx^3$$

is the same in all inertial reference systems; since $dx^i (i = 0, 1, 2, 3)$ are the components of a four-dimensional vector, the collection of quantities

$$\nabla_i \varphi = \frac{\partial \varphi}{\partial x^i} = \left(\frac{\partial \varphi}{\partial x^0}, \frac{\partial \varphi}{\partial x^1}, \frac{\partial \varphi}{\partial x^2}, \frac{\partial \varphi}{\partial x^3} \right)$$

is a four-dimensional vector. Thus, the operator of four-dimensional gradient defined in the form

$$\nabla_i = \left(\frac{\partial}{\partial x^0}, \nabla \right),$$

where ∇ is the operator of three-dimensional gradient, is transformed as a four-dimensional vector.

2.35. Compose a four-dimensional tensor T^{ik} from the partial derivatives $\partial A_i / \partial x^k$ ($i = 0, 1, 2, 3$), where A^i is a four-dimensional vector. Show that the four-dimensional divergence $\partial A^i / \partial x^i$ is an invariant under Lorentz transformations.

Answer.

$$T_{ik} = \nabla_k A_i, \quad \nabla_k = \left(\frac{\partial}{\partial x^0}, \frac{\partial}{\partial x^1}, \frac{\partial}{\partial x^2}, \frac{\partial}{\partial x^3} \right).$$

The four-dimensional divergence

$$\nabla_i A^i = \frac{\partial A^0}{\partial x^0} + \frac{\partial A^1}{\partial x^1} + \frac{\partial A^2}{\partial x^2} + \frac{\partial A^3}{\partial x^3} = \text{inv}.$$

2.36. Find the law of transformation of the quantities:

(a) $A^i A_i$; (b) $T_{ik} A^k$, if A^k is a four-dimensional vector, and T^{ik} is a four-dimensional tensor.

Answer. (a) scalar; (b) four-dimensional vector.

2.37. Write the Lorentz transformation (2.1) in the variables x^1, x^2, x^3, $x^0 = ct$. Present the relative velocity V in terms of the angle α by the formula $V/c = \text{th } \alpha$.

Answer. If $x^i = \alpha^i_k x'^k$, then the matrix $\hat{\alpha}$ takes the form

$$\hat{\alpha} = \begin{pmatrix} \text{ch}\,\alpha & -\text{sh}\,\alpha & 0 & 0 \\ \text{sh}\,\alpha & -\text{ch}\,\alpha & 0 & 0 \\ 0 & 0 & -1 & 0 \\ 0 & 0 & 0 & -1 \end{pmatrix}.$$

2.3. Relativistic electrodynamics

We now write the basic formulas of the relativistic electrodynamics in vacuum. The density of a three-dimensional current $\vec{j} = \rho \vec{v}$ and the charge density ρ form *the four-dimensional current density vector*

$$j^i = (c\rho, \vec{j}).$$

Electric and magnetic fields are components of the antisymmetric *four-dimensional electromagnetic field tensor* F^{ik}:

$$F^{ik} = \begin{pmatrix} 0 & -E_x & -E_y & -E_z \\ E_x & 0 & -H_z & H_y \\ E_y & H_z & 0 & -H_x \\ E_z & -H_y & H_x & 0 \end{pmatrix}.$$

Under the transition from the system S to the system S', the components of the fields are transformed by the formulas (the axis x and x' are parallel to the direction of the relative velocity):

$$E_x = E'_x, \quad E_y = \gamma^{-1}(E'_y + \beta H'_z), \quad E_z = \gamma^{-1}(E'_z - \beta H'_y);$$
$$H_x = H'_x, \quad H_y = \gamma^{-1}(H'_y - \beta E'_z), \quad H_z = \gamma^{-1}(H'_z + \beta E'_y).$$

The quantities

$$H^2 - E^2 = \text{inv}, \quad \vec{E} \cdot \vec{H} = \text{inv} \qquad (2.10)$$

are invariants under Lorentz transformations. The vector, $\vec{A}$, and scalar, φ, potentials form *the four-dimensional vector potential*

$$A^i = (\varphi, \vec{A}).$$

The components of the *tensor of energy-momentum* in vacuum are given by the formula

$$T^{ik} = \frac{1}{4\pi}\left(-F^{il}F_l^k + \frac{1}{4}g^{ik}F_{lm}F^{lm}\right).$$

Nine spatial components of the tensor T^{ik} form the three-dimensional *Maxwell tensor of strengths*

$$\sigma_{\alpha\beta} = \frac{1}{4\pi}\left\{(E_\alpha E_\beta + H_\alpha H_\beta) - \frac{1}{2}\delta_{\alpha\beta}(E^2 + H^2)\right\}.$$

The space-time components of T^{ik} are proportional to components of the energy flow density $\vec{S}$ and the field momentum density $\vec{g}$:

$$T_{0\alpha} = \frac{1}{c}S_\alpha, \quad \vec{S} = \frac{c}{4\pi}\vec{E} \times \vec{H}, \quad T_{0\alpha} = cg_\alpha, \quad \vec{g} = \frac{1}{4\pi c}\vec{E} \times \vec{H} = \frac{1}{c^2}\vec{S}.$$

The time component of T^{ik} is connected with the field energy density w by the relation

$$T_{00} = w = (1/8\pi)(E^2 + H^2).$$

The divergence of the tensor T^{ik} defines the bulk density of forces $f^i = (\vec{v} \cdot \vec{f}/c, \vec{f})$ acting on the charges:

$$\frac{\partial T_i^k}{\partial x^k} = f_i = \frac{1}{c}F_{ik}j^k.$$

We now consider the formulas of electrodynamics in the presence of media. In this case, the vectors of the fields $\vec{E}, \vec{D}, \vec{B}, \vec{H}$ form two antisymmetric four-dimensional second-rank tensors: the field tensor

$$F^{ik} = \begin{pmatrix} 0 & -E_x & -E_y & -E_z \\ E_x & 0 & -B_z & B_y \\ E_y & B_z & 0 & -B_x \\ E_z & -B_y & B_x & 0 \end{pmatrix}$$

and the induction tensor

$$H^{ik} = \begin{pmatrix} 0 & -D_x & -D_y & -D_z \\ D_x & 0 & -H_z & H_y \\ D_y & H_z & 0 & -H_x \\ D_z & -H_y & H_x & 0 \end{pmatrix}.$$

The vectors of polarization and magnetization $\vec{P}$ and $\vec{M}$ form also the four-dimensional tensor

$$M^{ik} = \begin{pmatrix} 0 & -P_x & -P_y & -P_z \\ P_x & 0 & -M_z & M_y \\ P_y & M_z & 0 & -M_x \\ P_z & -M_y & M_x & 0 \end{pmatrix}.$$

The formulas $\vec{D} = \vec{E} + 4\pi\vec{P}$ and $\vec{B} = \vec{H} + 4\pi\vec{M}$ can be joined in a single relation

$$H^{ik} = F^{ik} - 4\pi M^{ik}.$$

The four-dimensional force f^i acting on a unit volume from the side of a field is defined as

$$f^i = \left(\frac{1}{c}(Q + \vec{f} \cdot \vec{v}), \vec{f}\right),$$

where $\vec{f}$ is the ponderomotive force applied to a unit volume; Q is the Joule–Lenz heat released for a unit time in a unit volume.

* * *

Example 2.4. Let a homogeneous electromagnetic field with $\vec{E}$ and $\vec{H}$ be created in the reference system S. With which velocity relative to S must the system S', where $\vec{E}' \| \vec{H}'$, move? Does this problem have a solution and is it unique? What are the absolute values of $\vec{E}'$ and $\vec{H}'$?

Solution. The problem has an infinite number of solutions. If the system S' (which moves with velocity $\vec{V}$), where $\vec{E}' \| \vec{H}'$, is found, then $\vec{E}$ and $\vec{H}$ will be parallel in any reference system which moves relative S' along this common direction, as it follows from Eq. (2.10). Therefore, we will seek only a reference system S' that moves perpendicularly to the plane $(\vec{E}, \vec{H})$.

The parallelism of the vectors $\vec{E}'$ and $\vec{H}'$ implies that $\vec{E}' \times \vec{H}' = 0$;

$$\begin{cases} \vec{E}' = \vec{E} + \dfrac{1}{c}[\vec{V} \times \vec{H}], \\[2mm] \vec{H}' = \vec{H} - \dfrac{1}{c}[\vec{V} \times \vec{E}]; \end{cases}$$

$$\vec{E}' \times \vec{H}' = \left(\vec{E} + \frac{1}{c}[\vec{V} \times \vec{H}]\right) \times \left(\vec{H} - \frac{1}{c}[\vec{V} \times \vec{E}]\right)$$

$$= \vec{E} \times \vec{H} - \vec{E} \times \frac{1}{c}[\vec{V} \times \vec{E}] + \frac{1}{c}[\vec{V} \times \vec{H}]$$

$$-\frac{1}{c^2}[\vec{V} \times \vec{H}] \times [\vec{V} \times \vec{E}]$$

$$= [\vec{E} \times \vec{H}] - \frac{\vec{V}}{c}(E^2 - H^2) + [\vec{H} \times \vec{E}]\frac{V^2}{c^2}.$$

We have the square equation for $\frac{\vec{V}}{c}$:

$$[\vec{H} \times \vec{E}]\frac{V^2}{c^2} - (E^2 - H^2)\frac{\vec{V}}{c} + [\vec{E} \times \vec{H}] = 0;$$

$$D = (E^2 - H^2)^2 - 4(\vec{E} \cdot \vec{H});$$

$$\left(\frac{\vec{V}}{c}\right)_{1,2} = \frac{(E^2 - H^2) \pm \sqrt{(E^2 - H^2)^2 - 4(\vec{E} \cdot \vec{H})}}{2[\vec{E} \times \vec{H}]}.$$

Since $\frac{V}{c} \leq 1$, we take the solution

$$\frac{\vec{V}}{c} = \frac{(E^2 - H^2) \pm \sqrt{(E^2 - H^2)^2 - 4(\vec{E} \cdot \vec{H})}}{2[\vec{E} \times \vec{H}]}.$$

With the help of the invariants of the field, we obtain

$$H^2 - E^2 = H'^2 - E'^2;$$

$$E'^2 = \frac{1}{2}[E^2 - H^2 + \sqrt{(E^2 - H^2)^2 + 4(\vec{E} \cdot \vec{H})^2}];$$

$$H'^2 = \frac{1}{2}[H^2 - E^2 + \sqrt{(E^2 - H^2)^2 + 4(\vec{E} \cdot \vec{H})^2}].$$

Problems

2.38. Find the law of transformation of components of the electric $\vec{E}$ and magnetic $\vec{H}$ fields in vacuum in a system S', which moves relative to the system S with an arbitrarily directed velocity $\vec{V}$.

Answer. In vacuum:

$$\vec{E} = \gamma\left(\vec{E}' - \frac{\vec{V}}{c} \times \vec{H}'\right) - (\gamma - 1)\vec{V}\frac{(\vec{V} \cdot \vec{E}')}{V^2},$$

$$\vec{H} = \gamma\left(\vec{H}' - \frac{\vec{V}}{c} \times \vec{E}'\right) - (\gamma - 1)\vec{V}\frac{(\vec{V} \cdot \vec{H}')}{V^2}.$$

2.39. Show that the quantity $\vec{E}^2 - \vec{H}^2$ is invariant under the Lorentz transformations.

2.40. Prove that if the directions of the electric $\vec{E}$ and magnetic $\vec{H}$ fields are perpendicular to each other in some reference system, then they are perpendicular in all other inertial reference systems.

2.41. Let the electric $\vec{E}$ and magnetic $\vec{H}$ fields be mutually perpendicular in a reference system S. Which velocity relative to S should the system S', where there exists only the electric or only magnetic field, move with? Does the problem have a solution and is it unique?

2.42. Show that the wave equation is not invariant under the Galilei transformations and is invariant under the Lorentz transformations.

2.43. Find the law of a relativistic transformation of the Joule–Lenz heat Q, by starting from the definition of the four-dimensional force density.

Answer. Let u^i be the four-dimensional velocity of the medium. We can construct the four-dimensional invariant

$$-f_i u^i = \gamma(fv) - \gamma(Q + fv) = -\gamma Q = \text{inv}.$$

If we denote, by Q_0, the amount of heat which is released in a unit volume of the medium in a unit time in the system, where the medium is at rest, then $Q = Q_0\sqrt{1 - \beta^2}$.

2.44. Find the formulas of transformation for components of the tensor of energy-momentum T^{ik} under the Lorentz transformation.

Answer.

$$\omega = \gamma^2 \left(\omega' + \frac{2\beta}{c} S'_x + \beta^2 T'_{xx} \right),$$

$$S_x = \gamma^2[(1 + \beta^2)S'_x + V\omega' + VT'_{xx}];$$

$$S_y = \gamma(S'_y + VT'_{xy}), \quad S_z = \gamma(S'_z + VT'_{xz}),$$

$$T_{xx} = \gamma^2 \left(T'_{xx} + \frac{2\beta}{c} S'_x + \beta^2 \omega' \right);$$

$$T_{yy} = T'_{yy}, T_{yz} = T'_{yz}, T_{zz} = T';$$

$$T_{xy} = \gamma\left(T'_{xy} + \frac{\beta}{c}S'_y\right), \quad T_{xz} = \gamma\left(T'_{xz} + \frac{\beta}{c}S'_z\right).$$

2.45. Find the trace (sum of diagonal elements of a matrix) of the tensor of energy-momentum T^{ik}.

Answer. $T_i^i = 0$.

Section 3

Relativistic Mechanics

3.1. Energy and momentum

The momentum $\vec{p}$ of a relativistic particle is connected with its velocity $\vec{v}$ by the relation:

$$\vec{p} = \frac{m\vec{v}}{\sqrt{1 - v^2/c^2}},$$

where m is the rest mass of a particle.

The total energy of a particle, which moves freely, can be presented in terms of the velocity:

$$\varepsilon = \frac{mc^2}{\sqrt{1 - v^2/c^2}}$$

or the momentum

$$\varepsilon = c\sqrt{p^2 + m^2c^2}.$$

The kinetic energy T of a particle differs from the total energy by the rest energy $E_0 = mc^2$:

$$T = \varepsilon - mc^2.$$

The energy, momentum, and velocity of a particle are connected by the formula

$$\varepsilon\,\vec{v} = c^2\vec{p}.$$

The energy and the momentum of a particle are the time and space components of the four-dimensional vector of energy-momentum (four-dimensional momentum)

$$p^i = (\varepsilon/c, \vec{p}).$$

Under the transition from one inertial reference system to another one, the energy and the momentum are transformed by formulas (2.2). The square of the four-dimensional momentum is a relativistic invariant:

$$p^{i2} = \varepsilon^2/c^2 - p^2 = m^2 c^2.$$

A particle is called nonrelativistic if its kinetic energy is low, and ultrarelativistic if its kinetic energy is high, as compared with the rest energy. The velocity of an ultrarelativistic particle is close to the velocity of light, and its momentum is connected with the energy by the relation:

$$\varepsilon = cp.$$

Particles with zero mass and zero rest energy (photons, neutrinos) are always ultrarelativistic, and their velocities are exactly equal to c. The energy and the momentum of a photon in vacuum are connected with its frequency by the formulas:

$$\varepsilon = \hbar\omega, \quad p = \hbar\omega/c = \hbar k,$$

where $\hbar = 1.05 \cdot 10^{-27}$ erg·s is the Planck constant.

The total energy and momentum of a closed system of particles are conserved. This implies that if the particles do not interact with one another before the start and after the termination of some reaction (decay or collision), then the total four-dimensional momenta in the initial and final states are the same:

$$\sum_a p_a^{i(0)} = \sum_b p_b^i,$$

where the summation is carried out over all particles that are present before and after the reaction.

It is convenient to consider the collisions between particles in one of two reference systems: the laboratory system S or the center-of-mass system S'. In the latter, the total momentum $\vec{p}$ of

all particles of the system is zero. It is worth noting that it is sometimes useful to use the invariance of the squares of four-dimensional
momenta.

Two types of collisions are distinguished: elastic collisions, in
which the internal states and, hence, the masses of particles are
not changed, and inelastic collisions, in which the internal energies
(masses) of colliding particles vary, i.e., some old particles disappear
or some new particles are created. In the inelastic collision of two
particles, the sum of masses, $m_1 + m_2$, of colliding particles differs
from the sum M_k of the masses of particles after the collision by
the value

$$\Delta M = m_1 + m_2 - M_k,$$

which is called *the mass defect*.

The quantity $Q = c^2 \Delta M$ is called the *energy yield of the reaction*.
The reactions that are running by the scheme

$$a + b \rightarrow c + d,$$

where two particles are transformed into two other particles, are
called *two-particle* (a special case of a two-particle reaction is the
elastic scattering of two particles).

It is convenient to describe the kinematics of two-particle reactions with the help of the invariant variables s, t, u:

$$s = (p_a^i + p_b^i)^2, \quad t = (p_a^i - p_c^i)^2, \quad u = (p_a^i - p_d^i)^2,$$

where p_a^i, p_b^i, p_c^i are the four-dimensional momenta of particles, which
participate in the reaction.

Any of the quantities s, t, u can be presented in terms of two
other ones with the help of the relation

$$s + t + u = (m_a^2 + m_b^2 + m_c^2 + m_d^2)c^2.$$

The clear idea of the kinematics of a two-particle reaction is given
by the kinematic plane, on which the values of the variables s and t
(or s, t, and u) are drawn.

The laws of conservation of the energy and the momentum separate the domain of values of s, t, u on the kinematic plane, in which this reaction has a physical meaning.

Many formulas of relativistic kinematics take a simpler form if we use the system of units in which the speed of light $c = 1$. In this case, the mass, energy, and momentum are measured in the same units, for example, in MeV (1 MeV $= 10^6$ eV $= 10^{-3}$ GeV $= 1.602$ 10^{-6} erg). In a number of cases, the masses of elementary particles are measured in units of electron mass m_e (i.e., the system of units in which $m_e = 1$ is used).

The binding energy is the quantity:

$$B = \Delta M c^2 = \sum \varepsilon_{0n} - \varepsilon_{0nucleus},$$

where ε_{0n} is the rest energy of a nucleon, and $\varepsilon_{0nucleus}$ is the rest energy of a nucleus.

* * *

Example 3.1. Let the flux of monochromatic μ-mesons, which were created in upper layers of the atmosphere, fall vertically downward. Find the ratio of the intensities of the flux of μ-mesons at an altitude h above the sea level (I_h) and at sea level (I_0), by considering that the flux decreases in a layer of air of thickness h only due to the natural decay of μ-mesons. The energy of μ-mesons $\varepsilon = 4.2 \cdot 10^8$ eV, $h = 3$ km, and the mean intrinsic lifetime of a μ-meson $\tau_0 = 2.2 \cdot 10^{-6}$ s.

Solution. The ratio of the intensities

$$\frac{I_h}{I_0} = \exp \frac{h}{v\tau};$$

$\tau = \frac{\tau_0}{\sqrt{1-v^2/c^2}}$ is the lifetime of a μ-meson which moves with velocity v;

$$\frac{I_h}{I_0} = \exp \frac{h\sqrt{1 - \frac{V^2}{c^2}}mc^2}{v\tau_0 mc^2} = \frac{h}{c\tau_0}\frac{mc^2}{E} \approx 2.5.$$

If there were no relativistic transformation of the time, then the ratio of the intensities (we assume that the velocity of mesons is equal to c)

$$\frac{I'_h}{I'_0} \approx \exp\frac{h}{\tau_0 c} \approx 94.4.$$

Observations confirm the first result ($I_h/I_0 \approx 2.5$) and, therefore, give direct experimental proof of the existence of the relativistic effect of a deceleration of the course of a moving clock.

Example 3.2. Find the dependence of the energy of a γ-quantum, which is created as a result of the decay of a π^0-meson, on the angle ϑ between the directions of propagation of the quantum and the π-meson.

Determine the energy spectrum of the γ-quanta from the decay in the laboratory reference system.

Hint. The laws of conservation of the energy and the momentum imply that, in the rest system of a π^0-meson, the energy of the γ-quantum $\varepsilon' = mc^2/2$ (m is the mass of a π^0-meson).

Solution. Since the momentum of a photon $p = \varepsilon/c$, we have

$$\varepsilon = \frac{\varepsilon'}{\gamma(1 - \beta\cos\vartheta)}, \quad \varepsilon' = \frac{mc^2}{2}, \quad \beta = \frac{v}{c}.$$

Comparing the relation following from $d\varepsilon = -\frac{\varepsilon' d(1-\beta\cos\vartheta)}{\gamma(1-\beta\cos\vartheta)^2}$ with the angular distribution of γ-quanta from the decay, we get the distribution of probabilities for the energies of photons from the decay:

$$dW(\varepsilon) = \frac{|d\varepsilon|}{\varepsilon_{\max} - \varepsilon_{\min}},$$

where $\varepsilon_{\min} = \varepsilon'\sqrt{\frac{1-\beta}{1+\beta}}$ is the minimum value of the energy of a γ-quantum from the decay (if $\vartheta = \pi$), and $\varepsilon_{\max} = \varepsilon'\sqrt{\frac{1+\beta}{1-\beta}}$ is the maximum value of the energy of a γ-quantum from the decay (if $\vartheta = 0$). From this, it is seen that the spectrum of γ-quanta from the decay has a rectangular shape in the laboratory reference system, i.e., all values of the energy in the interval from $\varepsilon_{\min}$ to $\varepsilon_{\max}$ are equiprobable.

Example 3.3. Let a free excited nucleus at rest (the excitation energy equals $\Delta\varepsilon$) emit a γ-quantum. Find its frequency ω, if the mass of the excited nucleus is equal to m. What is the reason for $\omega \neq \Delta E/\hbar$? How will the result be changed, if the nucleus is rigidly fixed in a crystal lattice (Mössbauer effect)?

Solution. For the nucleus emitting a γ-quantum, we write the laws of conservation of energy and momentum:

$$\varepsilon + \Delta\varepsilon = \hbar\omega + \varepsilon + \varepsilon_\beta,$$

where $\varepsilon + \Delta\varepsilon$ is the energy of the free excited nucleus; $\hbar\omega$ is the energy of a γ-quantum; $\varepsilon + \varepsilon_\beta$ is the energy of the nucleus after the emission including the recoil energy:

$$0 = p_\gamma - p_\beta;$$

$$p_\beta = p_\gamma = \frac{\Delta\varepsilon}{c};$$

$$\varepsilon_\beta = \frac{p_\beta^2}{2m} = \frac{\Delta\varepsilon^2}{2mc^2};$$

$$\hbar\omega = \Delta\varepsilon - \varepsilon_\beta = \Delta\varepsilon - \frac{\Delta\varepsilon^2}{2mc^2} = \Delta\varepsilon\left(1 - \frac{\Delta\varepsilon}{2mc^2}\right);$$

$$\omega = \frac{\Delta\varepsilon}{\hbar}\left(1 - \frac{\Delta\varepsilon}{2mc^2}\right).$$

The energy $\hbar\omega$, which is borne by a γ-quantum, is less than $\Delta\varepsilon$ by the value $(\Delta\varepsilon)^2/(2mc^2)$, which is borne by the recoiling nucleus.

Under conditions of tough binding of the nucleus with the crystal lattice, the latter receives no energy (since its mass $M \gg m$ is very high), and the γ-quantum bears the whole energy, $\hbar\omega = \Delta\varepsilon$.

Example 3.4. A particle with mass m_1 and energy ε_0 undergoes elastic collision with a stationary particle with mass m_2. Present the collision angles ϑ_1, ϑ_2 of the particles in the laboratory reference system in terms of their energies ε_1, ε_2 after the collision.

Solution. By the law of conservation of the four-dimensional momentum,

$$p_1^{i(0)} + p_2^{i(0)} = p_1^i + p_2^i. \tag{3.1}$$

In order to determine the scattering angle of the first particle, we transfer p_{1i} on the left-hand side of Eq. (3.1) and square both sides:

$$p_1^{i(0)^2} + p_2^{i(0)^2} + p_1^{i2} + 2p_1^{i(0)}p_2^{i(0)} - 2p_1^{i(0)}p_1^i - 2p_2^{i(0)}p_1^i = p_2^{i2}.$$

As is known, $p_1^{i(0)^2} = p_1^{i2} = m_1^2 c^2$, $p_2^{i(0)^2} = p_2^{i2} = m_2^2 c^2$. The scalar products are transformed in the following way ($\vec{p}_2^{(0)} = 0$):

$$-p_1^{i(0)}p_2^{i(0)} = \vec{p}_1^{(0)} \cdot \vec{p}_2^{(0)} - \frac{1}{c^2}\varepsilon_1^{(0)}\varepsilon_2^{(0)} = -\varepsilon_0 m_2, \ p_2^{i(0)}p_1^i = m_2\varepsilon_1,$$

$$-p_1^{i(0)}p_1^i = \vec{p}_1^{(0)} \cdot \vec{p}_1 - \frac{1}{c^2}\varepsilon_1^{(0)}\varepsilon_1 = p_0 p_1 \cos\vartheta_1 - \frac{\varepsilon_0\varepsilon_1}{c^2}, \tag{3.2}$$

where $p_0 = \frac{1}{c^2}\sqrt{\varepsilon_0^2 - m_1^2 c^4}$.
Substituting the formulas in equality (3.2), we have

$$\cos\vartheta_1 = \frac{\varepsilon_1(\varepsilon_0 + m_2 c^2) - \varepsilon_0 m_2 c^2 - m_1^2 c^4}{c^2 p_0 p_1}.$$

Analogously,

$$\cos\vartheta_2 = \frac{(\varepsilon_0 + m_2 c^2)(\varepsilon_2 - m_2 c^2)}{c^2 p_0 p_1}.$$

Problems

3.1. Find the law of transformation of the energy and components of the momentum of a particle when transiting to a system which moves with velocity v relative to the initial system.

3.2. Express the momentum p of a relativistic particle in terms of its kinetic energy T. Express the velocity v of the particle in terms of its momentum p.

   Answer.

$$p = \frac{1}{c}\sqrt{T(T + 2mc^2)}; \quad v = \frac{cp}{\sqrt{p^2 + m^2 c^2}}.$$

3.3. A particle with mass m has energy ε. Find the velocity v of a particle. Consider, in particular, the nonrelativistic and ultrarelativistic cases.

Answer. $\beta = \frac{v}{c} = \sqrt{1 - (\varepsilon_0/\varepsilon)^2}$, $\varepsilon_0 = mc^2$.

In the nonrelativistic case, $\beta \approx \sqrt{2T/\varepsilon_0}$, in the ultrarelativistic one, $\beta = 1 - \frac{1}{2}(\frac{\varepsilon_0}{\varepsilon})^2$.

3.4. Find the approximate formulas for the kinetic energy T of a particle whose mass is m: (a) in terms of its velocity v and (b) in terms of its momentum p to within v^4/c^4 and p^4/m^4c^4, respectively, if $v \ll c$.

Answer. (a) $T = \frac{1}{2}mv^2 + \frac{3}{8}m\frac{v^4}{c^2} + \cdots$, (b) $T = \frac{p^2}{2m} - \frac{1}{8}\frac{p^4}{m^3c^2} + \cdots$

3.5. Find the velocity v of a particle with mass m and charge e undergoing the action of an accelerating potential difference V (initial velocity is zero). Simplify the general formula in nonrelativistic and ultrarelativistic cases (take two terms of the expansion).

Answer.

$$v = \sqrt{\frac{2eV}{m}} \frac{1 + \frac{eV}{2mc^2}}{\left(1 + \frac{eV}{mc^2}\right)^2}.$$

In the case where $eV \ll mc^2$, the velocity $v = \sqrt{\frac{2eV}{m}}$ $(1 - \frac{3}{4}\frac{eV}{mc^2}) \ll c$.

In the case where $eV \gg mc^2$, the velocity $v = c[1 - \frac{1}{2}$ $(\frac{mc^2}{eV})^2] \approx c$.

3.6. At the outlet of an accelerator, we have a beam of charged particles with the kinetic energy T and with the current intensity I. Find the pressure F of the beam on the absorbing surface and the power W that is released on the surface. The mass of a particle is m, and its charge is e.

Answer. $F = \frac{I}{ce}\sqrt{T(T + 2mc^2)}$, $W = \frac{I}{e}T$.

3.7. Some body moves with relativistic velocity v through a gas containing N particles with mass m in a unit volume. The velocities of the particles are small. Find the pressure p of the gas on an element of body's surface, which is placed perpendicularly to its velocity, if the particles are elastically reflected from body's surface.

Answer. $p = \frac{2mv^2 N}{1-v^2/c^2}$.

3.8. The reference system S' moves with velocity $\vec{V}$ relative to the system S. A particle with mass m, which has energy ε' and velocity $\vec{v'}$ in S', moves at the angle ϑ' to the direction of $\vec{V}$. Find the angle ϑ between the momentum $\vec{p}$ of the particle and the direction $\vec{V}$ in the system S. Express the energy and the momentum of the particle in S in terms of ϑ', ε' or ϑ', v'.

Answer.

$$\tan\vartheta = \frac{1}{\gamma}\frac{p'\sin\vartheta'}{p'\cos\vartheta' + V\frac{\varepsilon'}{c^2}} = \frac{1}{\gamma}\frac{\sin\vartheta'}{\cos\vartheta' + \frac{V}{v'}},$$

where $\gamma = \frac{1}{\sqrt{1-V^2/c^2}}$; $\varepsilon = \gamma(\varepsilon' + p'V\cos\vartheta')$, p and p' are the momenta of the particle in the systems S and S', respectively.

In the ultrarelativistic case, the above-presented approximate formula can be used, if $\cos\frac{\vartheta'}{2} \gg \sqrt{|1 - \frac{V}{v'}|}$, where $v' = p'\frac{c^2}{\varepsilon'}$ is the velocity of the particle in S'. The energy in the ultrarelativistic case takes the form

$$\varepsilon \approx pc \approx 2\gamma\varepsilon'\cos^2\frac{\vartheta'}{2}.$$

3.9. A π^0-meson moves with velocity v and decays into two γ-quanta. Find the angular distribution $dW/d\Omega$ of the γ-quanta from the decay in the laboratory reference system, by considering that the distribution of γ-quanta from the decay is spherically symmetric in the rest system of the π^0-meson.

Answer. $dW = \frac{d\Omega}{4\pi\gamma^2(1-\beta\cos\vartheta)^2}$; $\int_{4\pi} dW = 1$, where $\beta = v/c$.

3.10. Determine the mass m of some particle which decays into two particles with masses m_1 and m_2, if the momenta p_1 and p_2 of the particles created as a result of the decay and the angle θ between the directions of their velocities are available from experiment.

Answer.

$$m^2 = m_1^2 + m_2^2 + 2\left[\sqrt{(p_1^2 + m_1^2)(p_2^2 + m_2^2)} - p_1 p_2 \cos\vartheta\right], \quad c = 1.$$

3.11. Determine the mass m_1 of some particle that is one of two particles created as a result of the decay of a particle, whose mass is m and the momentum p. The momentum p_2, mass m_2, and the angle θ_2 of departure of the second particle created in the decay are known.

Answer.

$$m_1^2 = m^2 + m_2^2 - 2\left[\sqrt{(p^2 + m^2)(p_2^2 + m_2^2)} - pp_2 \cos \vartheta_2\right], \quad c = 1.$$

3.12. A particle, whose mass is m_1 and velocity is v, collides with a stationary particle with mass m_2 and is absorbed by it. Find the mass m and the velocity V of the created particle.

Answer.

$$m^2 = \varepsilon^2 - p^2 = m_1^2 + m_2^2 + \frac{2m_1 m_2}{\sqrt{1 - v^2}},$$

$$V = \frac{p}{\varepsilon} = \frac{m_1 v}{m_1 + m_2\sqrt{1 - v^2}}, \quad c = 1.$$

3.13. A particle with mass m_1 strikes a stationary particle, whose mass is m_2. The reaction creates a number of particles with the total mass M. If $m_1 + m_2 < M$, then, for low kinetic energies of the first moving particle, the reaction does not occur, since it contradicts the law of conservation of the energy. Find the minimum kinetic energy of the first particle (energy threshold T_0 of the reaction), starting from which the reaction becomes possible.

Answer. $T_0 = \frac{c^2}{2m_1}(M - m_1 - m)(M + m_1 + m)$.

3.14. Show that the annihilation of an electron–positron pair with the emission of one photon contradicts the law of conservation of energy-momentum.

3.15. A particle with energy ε and mass m_1 strikes a particle with mass m_2 at rest. Find the velocity of the center-of-mass of particles relative to the laboratory reference system under such a collision.

Answer. $v = \frac{c\sqrt{\varepsilon^2 - m_1^2 c^4}}{\varepsilon + m_2 c^2}$.

3.16. A particle with mass M decays into two particles with masses m_1 and m_2. Find the energies of the created particles in the system of their center-of-mass.

3.17. A particle, which moves with velocity v, decays into two particles, whose energies in the system of the center-of-mass are E_1 and E_2. Find the relation between the angle at which the particles fly apart, and the energies of particles in the laboratory system.

3.18. An ultrarelativistic particle, whose mass is m and energy is ε_0, scatters elastically on a stationary nucleus with mass $M \gg m$. Determine the dependence of the final energy ε of the particle on the angle ϑ of its scattering.

Answer.

$$\varepsilon = \frac{\varepsilon_0}{1 + \frac{\varepsilon_0}{Mc^2}(1 - \cos \vartheta)}.$$

3.19. Determine the dependence of the frequency of a photon scattered on an electron at rest on the scattering angle (Compton effect).

3.20. Prove that the emission and absorption of light by a free electron in vacuum are impossible.

3.2. Motion of charged particles in an electromagnetic field

In the electromagnetic field $\vec{E}, \vec{H}$, a point particle with charge e which moves with velocity $\vec{v}$ undergoes the action of the Lorentz force:

$$\vec{F} = e\vec{E} + \frac{e}{c}\vec{v} \times \vec{H}.$$

Per unit time, the kinetic energy of the particle is changed by the value

$$\vec{F} \cdot \vec{v} = e\vec{E} \cdot \vec{v} = \dot{\varepsilon} = \frac{d\varepsilon}{dt},$$

where ε is the energy of the particle.

The magnetic field does no work on the particle, since the magnetic force is perpendicular to the velocity of the particle. From the quantities $\vec{F}$ and $\frac{d\varepsilon}{dt}$, we can construct the four-dimensional vector (vector of Minkowski force):

$$F^i = \left(\frac{\vec{F} \cdot \vec{v}}{c\sqrt{1 - v^2/c^2}}, \frac{\vec{F}}{\sqrt{1 - v^2/c^2}} \right).$$

The four-dimensional force is presented in terms of the tensor of electromagnetic field F^{ik}: $F^i = \frac{e}{c}F^{ik}u_k$, where u^k is the four-dimensional velocity of the particle.

The differential equations of motion of the particle in the four-dimensional form are as follows:

$$\frac{dp^i}{d\tau} = eF^i \quad \text{or} \quad mc\frac{du^i}{ds} = \frac{e}{c}F^{ik}u_k. \tag{3.2}$$

By projecting these equations onto the spatial and temporal axes, we get the equation of motion of the particle in three-dimensional form and the law of conservation of energy:

$$\dot{\vec{p}} = e\vec{E} + \frac{e}{c}\vec{v} \times \vec{H}, \quad T = e\vec{v} \cdot \vec{E}.$$

Here, $T = \varepsilon - mc^2$ is the kinetic energy of the particle; $\vec{p}$ is its momentum, and the dot stands for the differentiation with respect to the time t.

The Lagrangian of a charged particle in the electromagnetic field with potentials φ, $\vec{A}$ takes the form:

• in the relativistic case

$$L = -mc\sqrt{1 - v^2/c^2} - U;$$

• in the nonrelativistic case

$$L = \frac{mv^2}{2} - U,$$

where

$$U = -\frac{e}{c}\vec{A} \cdot \vec{v} + e\varphi.$$

The quantity U plays the role of the potential energy of interaction of the particle with the external field. The equation of motion of the particle can be written in Euler–Lagrange form

$$\frac{d}{dt}\frac{\partial L}{\partial \dot{q}^i} = \frac{\partial L}{\partial q^i},$$

where q^i, $\dot{q}^i$ are the generalized coordinates and the velocity.

The current arising during the rotational (orbital) motion of a point charged particle around some center is characterized by a magnetic moment

$$\vec{m} = \kappa \vec{L}, \tag{3.3}$$

where $\kappa = e/2mc$ is the gyromagnetic ratio; m is the mass of the particle; and $\vec{L} = \vec{r} \times m\vec{v}$ is its angular momentum.

In an external magnetic field $\vec{H}$, if a particle undergoes the action of the rotational moment $\vec{N} = \vec{m} \times \vec{H}$, its angular momentum $\vec{L}$ varies with the time by the law $d\vec{L}/dt = \vec{N}$. According to expression (3.3), the time dependence of the magnetic moment is determined by the equation

$$\frac{d\vec{m}}{dt} = \kappa \vec{m} \times \vec{H}. \tag{3.4}$$

In addition to the mechanical and magnetic moments related to the orbital motion, the microparticles also have the intrinsic (spin) mechanical $\vec{s}$ and magnetic $\vec{m}_0$ moments, which are directed in parallel or in antiparallel:

$$\vec{m}_0 = \kappa_0 \vec{s}.$$

For an electron, $\kappa_0 = e/mc < 0$, where e is the charge of an electron; m is its mass. The time variation of the moment $\vec{m}_0$ is described by Eq. (3.4), where κ should be replaced by κ_0 and $\vec{m}$ by $\vec{m}_0$.

A neutron has no electric charge; however, it has the spin moment $\vec{m}_0$. Due to quantum effects, this moment can be oriented in an external magnetic field $\vec{H}(\vec{r})$ in two ways: in the direction of the field or in the opposite direction, and the initial orientation is conserved if certain conditions are satisfied. In this case, the motion of neutrons with the magnetic moment oreinted in the direction of the field or in the opposite direction can be considered as the motion of classical particles in a force field with the potential energy

$$U = \mp m_0 H,$$

where $H = |\vec{H}(\vec{r})|$.

As usual, the energy U is very small. Therefore, the magnetic field affects mostly only the motion of very slow ("cold") neutrons.

* * *

Example 3.5. A particle with charge e and mass m moves with an arbitrary velocity in a homogeneous constant electric field $\vec{E}$. At the initial moment in time $t = 0$, the particle is located at the coordinate origin and has momentum $\vec{p}_0$. Determine the three-dimensional coordinates and the time t of a particle in the laboratory system as functions of its intrinsic time τ. Eliminating τ, express the three-dimensional coordinates of the particle as functions of t. Consider, in particular, nonrelativistic and ultrarelativistic cases.

Solution. Choose the axis $x \| e\vec{E}$. The differential equations of motion in the four-dimensional shape are as follows:

$$\frac{d^2 x}{d\tau^2} = \frac{|e|E}{mc}\frac{d(ct)}{d\tau}, \quad \frac{d^2 y}{d\tau^2} = 0, \quad \frac{d^2 z}{d\tau^2} = 0, \quad \frac{d^2(ct)}{d\tau^2} = \frac{|e|E}{mc}\frac{dx}{d\tau}.$$

Integrating this system with regard for the initial conditions

$$x = y = z = ct = 0, \quad \frac{dx}{d\tau} = \frac{p_0 x}{m}, \quad \frac{dy}{d\tau} = \frac{p_0 y}{m},$$

$$\frac{dz}{d\tau} = 0, \quad c\frac{dt}{d\tau} = \frac{\varepsilon_0}{mc}, \quad \text{if } \tau = 0, \quad \text{where } \varepsilon_0 = \sqrt{c^2 p_0^2 + m^2 c^4},$$

we find the parametric equations for the particle trajectory in the four-dimensional space:

$$x = \frac{\varepsilon_0}{|e|E}\left(\operatorname{ch}\frac{|e|E\tau}{mc} - 1\right) + \frac{cp_{0x}}{|e|E}\operatorname{sh}\frac{|e|E\tau}{mc},$$

$$y = \frac{p_{0y}\tau}{m}, \qquad z = 0,$$

$$ct = \frac{\varepsilon_0}{|e|E}\text{sh}\frac{|e|E\tau}{mc} + \frac{cp_{0x}}{|e|E}\left(\text{ch}\frac{|e|E\tau}{mc} - 1\right). \qquad (3.5)$$

Equation (3.5) yields

$$\tau = \frac{mc}{|e|E}\ln\frac{p_{0x} + |e|Et + \sqrt{(p_{0x} + |e|Et)^2 + m^2c^2 + p_{0y}^2}}{p_{0x} + \varepsilon_0/c}. \qquad (3.6)$$

Using expression (3.6) and eliminating sh and ch from Eqs. (3.5) and (3.6), we obtain the law of motion in the three-dimensional form:

$$x(t) = \frac{c}{|e|E}\left[\sqrt{(p_{0x} + |e|Et)^2 + m^2c^2 + p_{oy}^2} - \frac{\varepsilon_0}{c}\right];$$

$$y(t) = \frac{cp_{0y}}{|e|E}\ln\frac{p_{0x} + |e|Et + \sqrt{(p_{0x} + |e|Et)^2 + m^2c^2 + p_{oy}^2}}{p_{0x} + \varepsilon_0/c};$$

$$z(t) = 0.$$

If $p_0 \ll mc$ and $t \ll \frac{mc}{|e|E}$, the motion is nonrelativistic.

In this case, the formulas for x, y, z are transformed into the ordinary nonrelativistic formulas of uniformly accelerated motion:

$$x(t) = \frac{p_{0x}}{m}t + \frac{|e|E}{2m}t^2; \quad y(t) = \frac{p_{0y}}{m}t.$$

After sometime ($t \gg \frac{mc}{|e|E}$), the velocity of the particle approaches c (even if it was small at first). In this case,

$$x(t) = ct - \frac{mc^2}{|e|E}, \quad y(t) = \frac{cp_{0y}}{|e|E}\ln\frac{2|e|Et}{mc}$$

and the motion becomes uniform (with velocity c). The motion that occurs for $p_{0y} = 0$ is called *hyperbolic*.

Example 3.6. Find the differential equations of motion of a relativistic particle in an electromagnetic field, by using the Lagrangian in cylindrical coordinates.

Hint. Calculating the derivative with respect to the time in the Euler–Lagrange equations, and consider that this derivative is taken along the particle trajectory so that r, α, z should be considered as functions of the time.

Solution.

$$\frac{d}{dt}\left(\frac{m\dot{r}}{\sqrt{1-v^2/c^2}}\right) = \frac{mr\dot{\alpha}^2}{\sqrt{1-v^2/c^2}} + eEr + \frac{e}{c}(-H_\alpha\dot{z} + H_z r\dot{\alpha});$$

$$(3.7)$$

$$\frac{d}{dt}\left(\frac{mr^2\dot{\alpha}}{\sqrt{1-v^2/c^2}}\right) = e\left[E_\alpha + \frac{1}{c}(H_r\dot{z} - H_z\dot{r})\right]r; \qquad (3.8)$$

$$\frac{d}{dt}\left(\frac{m\dot{z}}{\sqrt{1-v^2/c^2}}\right) = e\left[E_z + \frac{1}{c}(H_\alpha\dot{r} + H_r r\dot{\alpha})\right]. \qquad (3.9)$$

Equations (3.7) and (3.9) look like ordinary equations of motion (but with a variable mass $\frac{m}{\sqrt{1-v^2/c^2}}$). In this case, the right-hand side of Eq. (3.7) includes the term $\frac{mr\dot{\alpha}^2}{\sqrt{1-v^2/c^2}}$, which is independent of the form of electromagnetic forces (centrifugal force). Equation (3.9) gives the time derivative of the momentum of the particle relative to the axis z in terms of the z-component of the moment of the Lorentz force.

Problems

3.21. Write the relativistic equation of motion of a particle under the action of the force $\vec{F}$. Express the momentum in terms of the velocity $\vec{v}$ of the particle. Consider the cases where the velocity (a) changes only in magnitude; (b) changes only in direction; (c) $v \ll c$.

Answer.

$$\frac{m}{(1-v^2/c^2)^{1/2}}\frac{d\vec{v}}{dt} + \frac{mv\vec{v}}{c^2(1-v^2/c^2)^{3/2}}\frac{dv}{dt} = \vec{F};$$

$$\frac{m}{(1-v^2/c^2)^{1/2}}\frac{d\vec{v}}{dt} + \frac{mv\vec{v}}{c^2(1-v^2/c^2)^{3/2}}\frac{dv}{dt} = \vec{F};$$

(a) $\dfrac{m}{(1 - v^2/c^2)^{3/2}} \dfrac{d\vec{v}}{dt} = \vec{F}, \quad \text{if } \vec{v} \| \vec{F};$

(b) $\dfrac{m}{(1 - v^2/c^2)^{1/2}} \dfrac{d\vec{v}}{dt} = \vec{F}, \quad \text{if } \vec{v} \perp \vec{F};$

(c) $m \dfrac{d\vec{v}}{dt} = \vec{F}.$

The quantities $\frac{m}{(1-v^2/c^2)^{3/2}}$ and $\frac{m}{(1-v^2/c^2)^{1/2}}$ are sometimes called the longitudinal and transverse masses, respectively.

3.22. Write the formula that connects the vectors of forces acting on a particle in the laboratory system $\vec{F}$ and in the system $\vec{F}'$, where the particle is at rest. The velocity of the particle is $\vec{v}$. **Answer**.

$$\vec{F} = \frac{1}{\gamma}\vec{F}' + \left(1 - \frac{1}{\gamma}\right)\frac{(\vec{v} \cdot \vec{F}')\vec{v}}{v^2};$$

$$\vec{F}' = \gamma\vec{F} - (\gamma - 1)\frac{(\vec{v} \cdot \vec{F})\vec{v}}{v^2},$$

where $\gamma = \frac{1}{(1-v^2/c^2)^{1/2}}$.

3.23. What force acts from the observation point in the instantly associated system on a body with mass m, which is in a rocket and is at rest relative to it, if the rocket moves with a relativistic velocity v over a circular orbit with radius R? **Answer**. $F = \gamma^2 \frac{mv^2}{R}$.

3.24. Find the trajectory of motion of a charged particle in a homogeneous electric field with strength $\vec{E}$. Consider the limiting case of small velocities.

3.25. Determine the path l of a relativistic charged particle, which has charge e, mass m, and initial energy ε, in a homogeneous decelerating electric field parallel to the initial velocity of the particle. **Answer**. $l = \frac{\varepsilon - mc^2}{eE}$.

3.26. Determine the law and the trajectory of motion of a charged particle in a homogeneous magnetic field with strength $\vec{H}$, if the particle was located at a point with radius-vector $\vec{r}_0$ at $t = 0$ and had momentum $\vec{p}_0$.

3.27. A nonrelativistic particle with charge e and mass m moves in a perpendicular homogeneous constant electric field $\vec{E} = (0, E_y, E_z)$ and magnetic field $\vec{H} = (0, 0, H)$. At $t = 0$, the particle was located at the coordinate origin and had a velocity $\vec{v} = (v_{0x}, 0, v_{0z})$. Plot the functions $x(t)$, $y(t)$, $z(t)$, and draw the possible trajectories of the particle.

Hint. Simplify the integration by introducing the variable $u = x + i\,y$.

3.28. A relativistic particle moves in a parallel homogeneous constant electric field $\vec{E}$ and magnetic field $\vec{H}$ ($\vec{E} \| \vec{H} \| z$). At $t = 0$, the particle was at the coordinate origin and has the momentum $\vec{p}_0 = (p_{0x}, 0, p_{0z})$. Determine the dependence of x, y, z, t on the intrinsic time τ of the particle.
Answer.

$$x = \frac{p_{0x}c}{eH}\sin kH\tau + \frac{p_{0y}c}{eH}(\cos kH\tau - 1);$$

$$y = \frac{p_{0x}c}{eH}(\cos kH\tau - 1) + \frac{p_{0y}c}{eH}\sin kH\tau;$$

$$z = \frac{\varepsilon_0}{eH}(\operatorname{ch} kE\tau - 1) + \frac{p_{0z}c}{eH}\operatorname{sh} kE\tau;$$

$$ct = \frac{p_{0z}c}{eE}(\operatorname{ch} kE\tau - 1) + \frac{\varepsilon_0}{eE}\operatorname{sh} kE\tau,$$

where $k = e/mc$.

3.29. Determine the law of motion of a particle in the mutually perpendicular homogeneous constant electric field $\vec{E}$ and magnetic field $\vec{H}$. Do this in two ways: (a) by using the Lorentz transformation and by considering that the motion of the particle is known only in the electric or only in the magnetic field, and (b) by integrating Eq. (3.2).

3.30. Find the differential equations of motion of a relativistic particle in the electromagnetic field, by using the Lagrangian in the cylindrical coordinates.

Answer.

$$\frac{d}{dt}\left(\frac{m\dot{r}}{\sqrt{1-v^2/c^2}}\right) = \frac{mr\dot{a}^2}{\sqrt{1-v^2/c^2}} + eE_r + \frac{e}{c}(-H_a\dot{z} + H_z r\dot{a});$$

$$(3.10)$$

$$\frac{d}{dt}\left(\frac{mr^2\dot{a}}{\sqrt{1-v^2/c^2}}\right) = e\left[E_a + \frac{1}{c}(H_r\dot{z} - H_z\dot{r})\right]r; \qquad (3.11)$$

$$\frac{d}{dt}\left(\frac{m\dot{z}}{\sqrt{1-v^2/c^2}}\right) = e\left[E_z + \frac{1}{c}(H_a\dot{r} - H_r r\dot{a})\right].$$

Equation (3.10) takes the form of the ordinary Newton's equations of motion (but with the variable mass $\frac{m}{\sqrt{1-v^2/c^2}}$). In this case, the right-hand side of this equation includes the term $\frac{mr\dot{a}^2}{\sqrt{1-v^2/c^2}}$, which is independent of the form of electromagnetic forces (centrifugal force). Equation (3.11) gives the time derivative of the momentum of the particle relative to the axis z in terms of the z-component of the moment of the Lorentz force.

3.31. The long direct cylindrical cathode with radius a, along which the uniformly distributed current I flows, emits electrons with zero initial velocity. These electrons move under the action of an accelerating potential U to the long anode, which is coaxial with the cathode and has radius b. What should the minimum value of the potential difference U_{cr} between the cathode and the anode be in order that the electrons arrive at the anode despite the action of the magnetic field of the current I? **Answer.** The potential difference U should be greater than

$$U_{cr} = \sqrt{\frac{4I^2}{c^2}\ln^2\frac{b}{a} + \frac{m^2c^4}{e^2}} - \frac{mc^2}{|e|}.$$

For $|e|U \ll mc^2$ (nonrelativistic electrons), the above formula gives $U_{cr} = \frac{2I^2|e|}{mc^4}\ln^2\frac{b}{a}$.

3.32. Let a current I flow along an infinitely long cylindrical conductor with radius a. From the conductor surface, an electron, whose initial velocity v_0 is directed in parallel to the conductor, breaks away. Find the largest distance b to which the electron can move from the conductor axis.

Answer.

$$b = a \exp \frac{p_0 c^2}{I|e|}, \quad \text{where } p_0 = \frac{m v_0}{\sqrt{1 - v_0^2/c^2}}.$$

Section 4

Constant Electric and Magnetic Fields in Vacuum

4.1. Constant electric field in vacuum

In this section, we present some problems of determination of the potential $\varphi(\vec{r})$ and the strength of the field $\vec{E}(\vec{r})$ by the given distribution of charges, which is characterized by the volume $\rho(\vec{r})$, surface $\sigma(\vec{r})$, or linear $k(\vec{r})$ densities. The distribution of point charges can be described by the volume density $\rho(\vec{r}) = \sum_i q_i\, \delta(\vec{r} - \vec{r}_i)$, where q_i is the value of the i-th charge, $\vec{r}_i$ is the radius-vector of the i-th charge; and $\delta(\vec{r} - \vec{r}_i)$ is the δ-function. The strength of the electric field satisfies the Maxwell equations

$$\operatorname{div} \vec{E} = 4\pi\rho, \quad \operatorname{rot} \vec{E} = 0. \tag{4.1}$$

The integral form of the first equation in (4.1) is frequently used (Gauss electrostatic theorem):

$$\oint_S E_n df = 4\pi q, \tag{4.2}$$

where f is some closed surface; q is the total charge inside this surface.

The strength and the potential of the electric field are connected by the relations:

$$\vec{E} = -\operatorname{grad}\varphi, \quad \varphi(\vec{r}) = \int_{\vec{r}}^{\vec{r}_0} \vec{E} \cdot d\vec{r}, \quad \varphi(\vec{r}_0) = 0. \tag{4.3}$$

The potential φ satisfies the Poisson equation

$$\Delta\varphi = -4\pi\rho.$$

The potential is continuous and bounded at all points of the space, at which there are no point charges, in particular, on a charged surface. The normal derivatives φ are discontinuous on a charged surface:

$$E_{2n} - E_{1n} = 4\pi\sigma \quad \text{or} \quad \frac{\partial\varphi_1}{\partial n} - \frac{\partial\varphi_1}{\partial n} = 4\pi\sigma.$$

The normal is directed from domain 1 to domain 2.

On the surface of a double electric layer with power τ,

$$\frac{\partial\varphi_1}{\partial n} = \frac{\partial\varphi_1}{\partial n}, \quad \varphi_2 - \varphi_1 = 4\pi\tau$$

(normal $\vec{n}$ is directed from the negatively charged surface of the layer to the positively charged one).

The principle of superposition allows one to find the potentials of complicated systems of charges, by summing the potentials of elementary charges

$$\varphi(\vec{r}) = \int \frac{\rho(\vec{r}')dV'}{|\vec{r}' - \vec{r}|}. \tag{4.4}$$

In the case of a surface or linear distribution of charges, the volume integral in formula (4.4) is replaced by the corresponding surface or linear integral. In the case of a system of point charges, it is replaced by the sum over charges. This note concerns also all below-presented formulas containing the integrals over a distribution of charges.

In the majority of cases, it is difficult to directly calculate integral (4.4). Therefore, the potential is frequently written in the form of a series, which is formed by the expansion of the integrand in powers of x/r or x'/r and its termwise integration. Such an expansion can be obtained in Cartesian and spherical coordinates.

Cartesian coordinates. If $r > a$ (a is the largest distance of charges of the system from pole O), then

$$\varphi(x,\ y,\ z) = \frac{q}{r} - p_\alpha \frac{\partial}{\partial x_\alpha} \frac{1}{r} + \frac{Q_{\alpha\beta}}{2!} \frac{\partial^2}{\partial x_\alpha \partial x_\beta} \frac{1}{r}$$

$$- \frac{Q_{\alpha\beta\gamma}}{3!} \frac{\partial^3}{\partial x_\alpha \partial x_\beta \partial x_\gamma} \frac{1}{r} \dots \tag{4.5}$$

The multipole moments q, p_α, $Q_{\alpha\beta}$ are given by the volume integrals:

$$q = \int \rho(\vec{r}')dV' \text{ is the complete charge of the system;}$$

$$p_\alpha = \int \rho(\vec{r}')x'_\alpha \, dV' \text{ are components of the dipole moment;}$$

$$Q_{\alpha\beta} = \int \rho(\vec{r}')x'_\alpha x'_\beta dV' \text{ are components of the quadrupole moment.}$$

Under a rotation of the coordinate system, the quantities q, p_α, $Q_{\alpha\beta}, \ldots$ are transformed, respectively, as a scalar, vector, second-rank tensor, etc. The second and third terms of potential (4.5) can be written as

$$\varphi^{(p)} = \frac{\vec{p} \cdot \vec{r}}{r^3},$$

where $\vec{p} = (p_x, p_y, p_z)$ is the vector of the dipole moment of the system,

$$\varphi^{(Q)} = \frac{1}{2r^5}[(3x^2 - r^2)Q_{xx} + (3y^2 - r^2)Q_{yy} + (3y^2 - r^2)Q_{zz}$$

$$+ 6xyQ_{xy} + 6xzQ_{xz} + 6yzQ_{yz}].$$

Spherical coordinates. Let us use the expansion of $|\vec{r} - \vec{r}'|^{-1}$. Substituting it in expression (4.4), we obtain, if $r > r'$,

$$\varphi(\vec{r}) = \sum_{l=0}^{\infty} \sum_{m=-l}^{l} \sqrt{\frac{4\pi}{2l+1}} \frac{Q_{lm}Y_{lm}(\vartheta, \alpha)}{r^{l+1}}, \quad (r > r'),$$

where Q_{lm} of the multipole moment of the order l, m;

$$Q_{lm} = \sqrt{\frac{4\pi}{2l+1}} \int \rho(\vec{r}')r'^l Y^*_{lm}(\vartheta', \alpha') \, dV'.$$

If $r' > r$, then r and r' change places, and we have

$$\varphi(\vec{r}) = \sum_{l=0}^{\infty} \sum_{m=-l}^{l} \sqrt{\frac{4\pi}{2l+1}} r^l Q'_{lm} Y_{lm}(\vartheta, \alpha)(r < r'),$$

where

$$Q'_{lm} = \sqrt{\frac{4\pi}{2l+1}} \int \frac{\rho(\vec{r}')}{r'^{l+1}} Y^*_{lm}(\vartheta', \alpha') \, dV'.$$

If the observation point $\vec{r}$ is contained inside the distribution of charges, then it is necessary to divide the domain of integration in formula (4.4) into two parts by a sphere with radius r centered at the pole O. For the integration over a domain inside the sphere, we need to use the expansion in $|\vec{r} - \vec{r}'|^{-1}$. By integrating over the exterior domain, we will use the formula of expansion with the change of r by r'.

The energy of the electrostatic field can be calculated by one of the formulas:

$$W = \frac{1}{8\pi} \int E^2 \, dV, \quad W = \frac{1}{2} \int \rho\varphi \, dV \tag{4.6}$$

(these formulas are equivalent, if the charges are located in a bounded domain of the space, and if the integration is carried out over the whole space).

The energy of interaction of two systems of charges 1 and 2 is determined by the formulas:

$$U = \int \rho_1(\vec{r}) \, \varphi_2(\vec{r}) \, dV = \int \frac{\rho_1(\vec{r}_1) \, \rho_2(\vec{r}_2)}{|\vec{r}_1 - \vec{r}_2|} \, dV_1 dV_2.$$

The generalized ponderomotive forces can be obtained by the differentiation of U or W with respect to the corresponding generalized coordinates a_i:

$$F_i = -\frac{\partial U}{\partial a_i} \quad \text{or} \quad F_i = -\frac{\partial W}{\partial a_i}.$$

The generalized force is positive, if it increases the corresponding coordinate.

* * *

Example 4.1. Find the potential and the strength of the electric field inside and outside a uniformly charged cylinder with radius R. The charge of a unit length of the cylinder is χ.

Solution. The simplest method of solution is to use the Gauss electrostatic theorem.

To integrate the Poisson equation, it is necessary to use the Laplace operator in a cylindrical coordinate system with regard for the fact that φ depends only on r due to the symmetry of the system.

Let us use the Gauss theorem (4.2). Inside the given cylinder, we take a surface in the form of a coaxial cylinder with radius r and height h. The flow of the vector of tension through such a surface is

$$\oint_S E_n \, df = E_1 \cdot 2\pi r h.$$

If the cylinder is uniformly charged over its bulk, then the Gauss surface contains a charge equal to the product of the volume charge density by the volume of the Gauss cylinder, i.e.,

$$q = \rho \pi r^2 h = \frac{Q}{\pi R^2 l} \pi r^2 h = \chi h \frac{r^2}{R^2}$$

(here, the linear charge density $\chi = \frac{Q}{l}$).

Then, by the Gauss theorem, $2\pi r h E_1 = 4\pi \chi h \frac{r^2}{R^2}$. From whence, we get that the electric field in the cylinder $E_1 = \frac{2\chi r}{R^2}$.

Relations (4.3) yield the potential

$$\varphi_1 = -\int \frac{2\chi r}{R^2} dr = -\frac{\chi r^2}{R^2} + C_1.$$

Choosing the zero potential on the cylinder surface, where $r = R$, we get that the constant $C_1 = \chi$.

Finally, we have

$$\varphi_1 = -\frac{\chi r^2}{R^2} + \chi = \chi \left(1 - \frac{r^2}{R^2}\right). \quad (r \leq R)$$

We now consider the Gauss surface outside the cylinder. It contains the charge

$$q = \rho \pi R^2 h = \frac{Q}{\pi R^2 l} \pi R^2 h = \chi h.$$

By the Gauss theorem, we have $2\pi r h E_2 = 4\pi\chi h$, and

$$E_2 = \frac{2\chi}{r}.$$

Outside the cylinder, the potential is as follows:

$$\varphi_2 = -\int \frac{2\chi}{r} dr = -2\chi \ln r + C_2.$$

The integration constant is again determined from the equality of the potential to zero on the cylinder surface: if $r = R$, $C_2 = 2\chi \ln R$. Hence,

$$\varphi_2 = -2\chi \ln r + 2\chi \ln R = 2\chi \ln \frac{R}{r} \quad (r \geq R).$$

In the case of a surface distribution of a charge over the cylinder

$$E_1 = 0, \quad \varphi_1 = C_1 = 0, \quad (r \leq R),$$

the values of the electric field and the potential outside the cylinder $(r \geq R)$ coincide with those for the volume distribution of a charge.

Example 4.2. Let the charge distribution be spherically symmetric: $\rho = \rho(r)$. By dividing the charged domain into spherical layers, express the potential φ and the strength of the field $\vec{E}$ in terms of $\rho(r)$ (write φ and $\vec{E}$ in the form of a single integral with respect to r).

Solution. We divide the whole charged domain into concentric infinitely thin layers dr' in thickness. Let such a layer be located at the distance r' from the center of symmetry (Fig. 4.1). First, we calculate the strength of the field δE and potential $\delta\varphi$, which are created by such a layer at some point r of the space. Then we integrate the obtained formulas over all such layers.

Let r be a point located outside the charged layer (Fig. 4.1, a). Using the corresponding Maxwell equation in integral form, we get the following relation for the field strength (we denote it by δE_1):

$$\int \vec{\delta E_1} \cdot \vec{df} = 4\pi \int \rho dV.$$

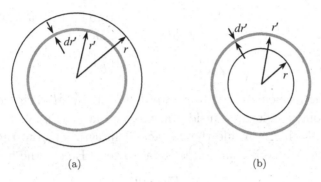

(a)   (b)

Fig. 4.1

We note that the symbol δ is used only in order to avoid confusion with the integration element, though $\delta E_1 \equiv dE_1$, in fact.

We take the sphere with radius r as the surface of integration, since the field intensity modulus is the same at every point of this sphere due to the symmetry of the problem. Therefore, the value of δE_1 can be taken outside the integral. In addition, the vectors $\overrightarrow{\delta E_1}$ and $\overrightarrow{df}$ are colinear. Hence, $\overrightarrow{\delta E_1} \cdot \overrightarrow{df} = \delta E_1 \cdot df$.

On the right-hand side of the equation, $\int \rho dV$ is the whole charge, which is contained in the sphere with radius r, i.e., the charge of a layer, which is equal to $\rho(r') \cdot 4\pi r'^2 dr'$ (since the layer is considered infinitely thin, we may assume that $\rho(r') = $ const). Thus, we have:

$$\delta E_1 \int df = 4\pi \rho(r') \cdot 4\pi r'^2 dr';$$

$$\delta E_1 \cdot 4\pi r^2 = 4\pi \rho(r') \cdot 4\pi r'^2 dr';$$

$$\delta E_1(r) = \frac{4\pi}{r^2}\rho(r')r'^2 dr'.$$

To calculate the corresponding potential, we take the relation

$$\overrightarrow{\delta E} = -\nabla \delta\varphi,$$

i.e.,

$$\delta\varphi_1(r) = -\int \overrightarrow{\delta E_1} \cdot \overrightarrow{dr} = -\int \delta E_1 \cdot dr$$

$$= -4\pi\rho(r')r'^2 dr' \cdot \int \frac{dr}{r^2} = \frac{4\pi}{r}\rho(r')r'^2 dr' + C_1.$$

The constant C_1 can be chosen with regard for the natural boundary condition $\delta\varphi_1|_\infty = 0$. It yields $C_1 = 0$. Hence,

$$\delta\varphi_1(r) = \frac{4\pi}{r}\rho(r')r'^2 dr'.$$

We now calculate the field (we denote it by δE_2), if the observation point r is located inside the charged layer (Fig. 4.1, b). Since there is no charge inside the surface of integration, the right-hand side of the Maxwell equation is equal to zero. Hence, the field

$$\delta E_2 = 0.$$

The corresponding potential in this domain is equal to a constant

$$\delta\varphi_2(r) = C_2.$$

We determine the constant C_2 from the condition of sewing of the potential on the boundary of the domains, i.e.,

$$\delta\varphi_2(r') = \delta\varphi_1(r').$$

This yields

$$C_2 = 4\pi\rho(r')r'dr',$$

and, hence,

$$\delta\varphi_2 = 4\pi\rho(r')r'dr'.$$

Now, when the formulas for the field and the potentials are found, it remains to integrate them over all values of r'. It should be remembered that the total field at every point r is composed of the fields of the layers placed relative to the observation point closer to the symmetry center (to determine this component of the potential, we need to integrate $\delta\varphi_1$ with respect to r' from zero to r) and those placed at distances larger than r (in this case, we integrate $\delta\varphi_2$ with respect to r' from r to infinity). Thus, we get the total potential:

$$\varphi(r) = \int_0^r \delta\varphi_1(r') + \int_r^\infty \delta\varphi_2(r')$$

$$= \frac{4\pi}{r}\int_0^r \rho(r')r'^2 dr' + 4\pi\int_r^\infty \rho(r')r'dr'.$$

Respectively, the strength of the field

$$E = \int_0^r \delta E_1(r') = \frac{4\pi}{r^2} \int_0^r \rho(r') r'^2 dr',$$

or, with regard for the direction of the field

$$\vec{E} = \frac{4\pi \vec{r}}{r^3} \int_0^r \rho(r') r'^2 dr'.$$

Example 4.3. Find the force $\vec{F}$ and the rotational moment $\vec{N}$ applied to the electric dipole with moment $\vec{p}$ in the field of a point charge q.

Solution. If we consider the dipole as a system of charges in the external field created by the charge q, then the formula for the total force acting on such a system can be written, to within linear terms, as

$$\vec{F} = \vec{E}_0 \sum e_\alpha + (\text{grad}\,(\vec{p} \cdot \vec{E}))_0.$$

The first term of this expression is zero, since it is the sum of charges of the dipole. Thus $\vec{F} = (\vec{p}\nabla)\vec{E}$.

Substituting the formula for the field strength of a point charge $\vec{E} = \frac{q}{R^3}\vec{R}$ in the formula for a force, we get

$$\vec{F} = q \left\{ \frac{1}{R^3}\,(\vec{p} \cdot \nabla)\vec{R} + \vec{R}\left(\vec{p} \cdot \nabla \frac{1}{R^3} \right) \right\}$$

$$= q \left\{ \frac{\vec{p}}{R^3} - \vec{R} \cdot \frac{3}{R^4}(\vec{p} \cdot \nabla R) \right\} = q \left\{ \frac{\vec{p}}{R^3} - \frac{3\vec{R}(\vec{p} \cdot \vec{R})}{R^5} \right\}$$

or, with regard for $\vec{n} = \frac{\vec{R}}{R}$,

$$\vec{F} = q\frac{\vec{p} - 3\vec{n}(\vec{p} \cdot \vec{n})}{R^3}.$$

The force acting on the dipole in the field of a point charge is equal by modulus and is opposite by direction to the force acting on the charge in the field of a dipole.

The moment of forces, which acts on the dipole in the external field $\vec{E}$, is $\vec{N} = \vec{p} \times \vec{E}$. Substituting the formula for the field strength of a point charge, we get

$$\vec{N} = q\frac{\vec{p} \times \vec{R}}{R^3}.$$

Problems

4.1. The infinite plane plate with thickness a is uniformly charged in bulk with density ρ. Find the potential φ and the strength $\vec{E}$ of the electric field.

Answer.

$$\varphi_1 = -2\pi\rho z^2, \quad \vec{E}_1 = 4\pi\rho z\vec{e}_z \left(|z| < \frac{a}{2}\right);$$

$$\varphi_2 = -\frac{1}{2}\pi\rho a(4|z| - a); \quad \vec{E}_2 = 2\pi\rho a\frac{z}{|z|}\vec{e}_z \left(|z| > \frac{a}{2}\right).$$

4.2. Determine the potential and the strength of the electric field inside and outside a uniformly charged ball. The volume charge density is ρ, and the ball radius is R.

Answer.

$$\varphi_1(r) = \frac{q}{R}\left(\frac{3}{2} - \frac{r^2}{2R^2}\right), \quad \vec{E}_1 = \frac{q\vec{r}}{R^3}(r \leq R);$$

$$\varphi_2(r) = \frac{q}{R}, \quad \vec{E}_2 = \frac{q\vec{r}}{r^3} \quad (r \geq R).$$

4.3. A uniformly charged ball with volume charge density ρ has a ball-like cavity, whose center is located at the distance a from the ball center. Find the strength of the electric field inside the cavity, inside the ball, and outside the ball. The radii of the ball and the cavity are, respectively, R and R'.

Answer.

$$\text{(a) } \vec{E} = \frac{4}{3}\pi\rho\vec{a}; \quad \text{(b) } \vec{E} = \frac{4}{3}\pi\rho\vec{r} - \frac{4}{3}\pi\rho\frac{R'^3}{|\vec{r} - \vec{a}|^3}(\vec{r} - \vec{a});$$

$$\text{(c) } \vec{E} = \frac{4}{3}\pi\rho\left[\frac{R^3}{r^3}\vec{r} - \frac{R'^3}{|\vec{r} - \vec{a}|^3}(\vec{r} - \vec{a})\right].$$

4.4. A layer of a dielectric with a volume charge density ρ is bounded by two parallel planes. The dielectric thickness is equal to d. Find the strengths of the electric field inside and outside the dielectric.
Answer.

$$\vec{E} = 2\pi\rho d\frac{\vec{z}}{|z|}(outside), \quad \vec{E} = 4\pi\rho z(inside).$$

4.5. Determine the electrostatic field energy for the charge distribution according to Problem 4.2 and Example 4.1. Perform the calculations in two ways [see formulas (4.6)].
Answer. $W = \dfrac{3q^2}{5R}$.

4.6. Determine the potential formed by the electron of a hydrogen atom, by considering that the electron charge in the ground state is distributed with the volume density $\rho = \frac{e}{\pi a^3}e^{-2r/a}$, where a is a constant.
Answer.

$$\varphi = \frac{e}{r}(1 - e^{-2r/a}) - \frac{e}{a}e^{-2r/a}.$$

4.7. Find the potential and the strength of the electric field on the axis of a uniformly charged planar ring with surface density σ (the inner and outer radii of the ring are R_1 and R_2). Consider the following special cases: (a) the field of a planar disk ($R_1 \to$ 0); (b) the field of a charged plane ($R_1 \to 0$, $R_2 \to \infty$).
Answer.

$$\varphi = 2\pi\sigma(\sqrt{R_2^2 + z^2} - \sqrt{R_1^2 + z^2}),$$

$$E_x = E_y = 0, \quad E_z = 2\pi\sigma\left(\frac{z}{\sqrt{R_1^2 + z^2}} - \frac{z}{\sqrt{R_2^2 + z^2}}\right).$$

The special cases:

$$(a)\ E_z = 2\pi\sigma\left(\frac{z}{|z|} - \frac{z}{\sqrt{R_2^2 + z^2}}\right), \quad (b)\ E_z = 2\pi\sigma\frac{z}{|z|}.$$

4.8. The ellipsoid with semiaxes a, b, c is uniformly charged over its bulk; and the total charge of the ellipsoid is q. Find the potential at large distances from such a system to the quadrupole term. Consider the separate cases of an ellipsoid of rotation with semiaxes $a = b$ and c and a ball $(a = b = c)$.
Answer.

$$\varphi(x, y, z) \approx \frac{q}{r} + q\frac{a^2(3x^2 - r^2) + b^2(3y^2 - r^2) + c^2(3z^2 - r^2)}{10r^5};$$

if $a = b$, then $\varphi(r, \theta) = \frac{q}{r} + q\frac{c^2-a^2}{5}\frac{P_2(\cos\theta)}{r^3}$;
if $a = b = c$, then $\varphi = \frac{q}{r}$.

4.9. Calculate the energy of interaction of the electron cloud of a hydrogen atom with the proton. Take the charge density of the electron cloud from Problem 4.6.
Answer. $W = \frac{-e^2}{a}$.

4.10. Calculate the energy of interaction of two balls, whose charges e_1 and e_2 are distributed spherically symmetrically. The distance between the centers of the balls is a, and the radii of the balls are R_1 and R_2, respectively.
Answer. $W = \frac{e_1e_2}{a}$.

4.11. A dipole with moment $\vec{p}_1$ is placed at the coordinate origin, and a dipole with moment $\vec{p}_2$ is located at a point with radius-vector $\vec{r}$. Find the energy of interaction U of these dipoles and the force $\vec{F}$ acting between them. At which orientation of dipoles will this force be maximum?
Answer.

$$U = p_1p_2\frac{\sin\vartheta_1\sin\vartheta_2\cos\varphi - 2\cos\vartheta_1\cos\vartheta_2}{r^3},$$

where ϑ_1 is the angle between r and p_1; $\vartheta_2\vartheta_1$ is the angle between r and p_2, and φ ϑ_1 is the angle between the planes (r, p_1) and (r, p_2),

$$F = 3p_1p_2\frac{\sin\vartheta_1\sin\vartheta_2\cos\varphi - 2\cos\vartheta_1\cos\vartheta_2}{r^4}.$$

4.2. Constant magnetic field

In the case of a constant magnetic field, the Maxwell equations take the form

$$\text{rot } \vec{H} = \frac{4\pi}{c} \vec{j}, \quad \text{div } \vec{B} = 0, \tag{4.7}$$

where $\vec{H}$ is the strength of the magnetic field; $\vec{j}$ is the bulk current density; c is the electrodynamic constant (light velocity); and $\vec{B}$ is the magnetic induction.

In the isotropic diamagnetics and paramagnetics, $\vec{B}$ and $\vec{H}$ are coupled by the relation

$$\vec{B} = \mu \vec{H}, \tag{4.8}$$

where μ is the magnetic permeability (scalar) of a substance; for the anisotropic substances, μ is a second-rank tensor.

The density of molecular currents $\vec{j}_{\text{mol}}$ in a substance that is in a constant magnetic field can be expressed in terms of the magnetization vector $\vec{M}$ (magnetic moment of a unit volume) by the formula

$$\vec{j}_{\text{mol}} = c \text{ rot } \vec{M}. \tag{4.9}$$

Vector $\vec{M}$ is connected with $\vec{B}$ and $\vec{H}$ by the relation

$$\vec{B} = \vec{H} + 4\pi \vec{M}.$$

The main methods of solution of the problem of determination of the magnetic field in a nonferromagnetic medium:

Use of the Biot–Savart law. A current element $I \, d\vec{l}$ creates the magnetic field

$$d\vec{H} = \frac{I}{cr^3} [d\vec{l} \times \vec{r}] \tag{4.10}$$

in vacuum or in a homogeneous medium. By the principle of superposition, the total field at a given point can be determined by the integration of (4.10) over all elements of the current (with respect to $d\vec{l}$).

Direct integration of the system of equations (4.8), (4.9) with the boundary conditions

$$\vec{n} \cdot (\vec{B}_2 - \vec{B}_1) = 0, \quad \vec{n} \times (\vec{H}_2 - \vec{H}_1) = \frac{4\pi}{c} \vec{i}, \qquad (4.11)$$

where $\vec{i}$ is the surface current density, and the normal $\vec{n}$ is directed from the first domain to the second one.

If the distribution of currents is axially symmetric, the integral form of the first equation in (4.7) can be used:

$$\oint \vec{H} \, d\bar{l} = \frac{4\pi}{c} I. \qquad (4.12)$$

Here, the integral is taken over any closed contour; I is the total current flowing through an arbitrary surface relying on this contour.

Method of vector potential. The vector potential $\vec{A}$ is determined by the relation

$$\vec{B} = \text{rot } \vec{A}$$

and the additional condition

$$\text{div } \vec{A} = 0.$$

In the domains, where a magnetic is homogeneous, $\vec{A}$ satisfies the equation

$$\Delta \vec{A} = -\frac{4\pi\mu}{c} \vec{j}.$$

The boundary conditions for the vector potential follow from the boundary conditions (4.11) for $\vec{B}$ and $\vec{H}$.

The vector potential formed by the given distribution of currents can be presented (in a homogeneous medium with magnetic permeability μ) in the form of an integral over the volume in which the current flows:

$$\vec{A}(\vec{r}) = \frac{\mu}{c} \int \frac{\vec{j}(\vec{r}')}{|\vec{r} - \vec{r}'|} \, dV'.$$

The corresponding formula for a linear current is obtained by the change $\vec{j} \, dV' \rightarrow I d\vec{l}'$. At large distances from the domain in which the current flows, Eq. (4.12) takes the form

$$\vec{A} = \frac{\vec{m} \times \vec{r}}{r^3},$$

where $\vec{m} = \frac{1}{2c} \int [\vec{r}' \times \vec{j}]\, dV'$ is the magnetic dipole moment (we set $\mu = 1$).

Method of scalar potential. In those spatial domains where $\vec{j} = 0$, we have rot $\vec{H} = 0$. Therefore, we may consider that

$$\vec{H} = -\text{grad}\ \psi,$$

where ψ is a scalar potential satisfying the Laplace equation for $\mu = $ const.

However, the scalar potential introduced in such way is not a univalent function of the observation point in the general case.

The energy of a magnetic field localized in some volume V is given by the integral over that volume:

$$W = \frac{1}{8\pi} \int (\vec{H} \cdot \vec{B})\, dV. \tag{4.13}$$

If the system of currents has bounded sizes, its total energy can be calculated by the formula

$$W = \frac{1}{2c} \int (\vec{A} \cdot \vec{j})\, dV,$$

in which the integration is executed over the volume where a current flows.

The magnetic energy of a quasilinear conductor with current I is expressed in terms of the coefficient of self-induction L of the conductor:

$$W = \frac{LI^2}{2c^2}. \tag{4.14}$$

The inductance can be also presented in terms of the double integral over the volume of the conductor:

$$L = \frac{1}{I} \iint \frac{\vec{j}(\vec{r})\vec{j}(\vec{r}')}{|\vec{r} - \vec{r}'|}\, dV\, dV'.$$

The energy of interaction of two conductors with current is described by the formulas

$$W = \frac{1}{4\pi} \int (\vec{H}_1 \cdot \vec{B}_2)\, dV = \frac{1}{c} \int (\vec{j}_1 \cdot \vec{A}_2)\, dV, \quad W_{12} = W_{21}.$$

The first integral is taken over the whole space, and the second one is calculated over the volume of one of the conductors. In the case of quasilinear currents, the energy can be presented in terms of the coefficient of mutual induction L_{12}:

$$W_{12} = c^{-2} L_{12} I_1 I_2. \tag{4.15}$$

Formula (4.15) can be presented in the form

$$W_{12} = I_1 \Phi_{12}/c^2,$$

where Φ_{12} is the magnetic induction flow formed by the second current through the contour of the first current:

$$\Phi_{12} = \int \vec{B}_2 \cdot d\vec{S}_1 = \oint \vec{A}_2 \cdot d\vec{l}_1 = \frac{1}{c} L_{12} I_2. \tag{4.16}$$

The coefficient of mutual induction can be calculated from the expression for energy (4.15) and the magnetic induction flow (4.16) or, in the case of linear currents, by the formula

$$L_{12} = \mu \oint \oint \frac{d\vec{l}_1 \cdot d\vec{l}_2}{r_{12}}.$$

The generalized forces F_i acting between two constant currents can be determined by the differentiation of the energy of interaction W_{12} (or the quantity $U_{12} = -W_{12}$, which is called the potential function) with respect to the corresponding generalized coordinates:

$$F_i = \frac{\partial W_{12}}{\partial q_i} = -\frac{\partial U_{12}}{\partial q_i}.$$

The forces can be calculated also by the Ampère formula:

$$d\vec{F} = \frac{I}{c}(d\vec{l} \cdot \vec{B}),$$

where $d\vec{l}$ is an element of the contour over which a current I flows; $d\vec{F}$ is the force acting on this element from the side of the external field $\vec{B}$.

* * *

Example 4.4. In the thin conducting cylindrical shell with radius b, a conductor with radius a coaxial to it is placed. Over these conductors, constant currents of the same intensity J flow in opposite directions. Determine the magnetic field $\vec{H}$ formed by such system at all points of space. Solve the problem in two ways: by the integration of the Maxwell differential equations and with the help of the Maxwell equation in integral form: $\oint H_l dl = \frac{4\pi}{c} J$.

Solution. **First method.** It is convenient to solve the problem in a cylindrical coordinate system. Let the direction of the current over the conductor with a smaller radius coincide with the direction of the axis z. The curl and the divergence of a vector $\vec{H}$ can be written in this system in the following way:

$$\text{rot } \vec{H} = \left(\frac{1}{r} \frac{\partial H_z}{\partial \alpha} - \frac{\partial H_\alpha}{\partial z} \right) \vec{e}_r + \left(\frac{\partial H_r}{\partial z} - \frac{\partial H_z}{\partial r} \right) \vec{e}_\alpha$$

$$+ \frac{1}{r} \left(\frac{\partial (r H_\alpha)}{\partial r} - \frac{\partial H_r}{\partial \alpha} \right) \vec{e}_z;$$

$$\text{div } \vec{H} = \frac{1}{r} \frac{\partial}{\partial r} (r H_r) + \frac{1}{r} \frac{\partial H_\alpha}{\partial \alpha} + \frac{\partial H_z}{\partial z}.$$

Since $\vec{j} = j \vec{e}_z$, we have $(\text{rot } H)_z = \frac{4\pi}{c} j$.

With regard for Eq. (4.7), $H_r = 0$, $H_z = 0$, $\frac{1}{r} \frac{\partial}{\partial r} (r H_\alpha) = \frac{4\pi}{c} j$.

In the domain $r < a$,

$$\frac{1}{r} \frac{\partial}{\partial r} (r H_\alpha) = \frac{4\pi}{c} \frac{J}{\pi a^2}.$$

From this, by the integration with respect to r, we get $H_\alpha = \frac{2Jr}{ca^2}$.

In the domain $a \leq r \leq b$,

$$\frac{1}{r} \frac{\partial}{\partial r} (r H_\alpha) = 0.$$

From this, $H_\alpha = \frac{C_1}{r}$, where C_1 is the integration constant, which can be determined from the boundary condition (4.11):

$$\frac{C_1}{a} - \frac{2Ja}{ca^2} = 0.$$

We get $C_1 = \frac{2J}{c}$, and the formula for the magnetic field takes the form

$$H_\alpha = \frac{2J}{cr}.$$

In the domain $r > b$,

$$\frac{1}{r}\frac{\partial}{\partial r}(rH_\alpha) = 0,$$

which yields $H_\alpha = \frac{C_2}{r}$. The integration constant C_2 can be determined also from the boundary condition (4.11):

$$\frac{C_2}{b} - \frac{2J}{cb} = -\frac{4\pi}{c}\frac{J}{2\pi b}.$$

From this, we have $C_2 = 0$ and $H_\alpha = 0$.

Second method. In order to execute the integration by formula (4.12), we express an element $\vec{dl}$ of the line in terms of the cylindrical coordinates r and α:

$$\vec{dl} = \vec{e}_\alpha r d\alpha;$$

$$\oint \vec{H}\, \vec{e}_\alpha r d\alpha = \int_0^{2\pi} H_\alpha r d\alpha = H_\alpha 2\pi r = \frac{4\pi}{c}I.$$

In the first domain, $j = \frac{J}{\pi a^2}$, $I(r) = j\pi r^2$. Then we have eventually:

$$H_\alpha 2\pi r = \frac{4\pi}{c}J\frac{r^2}{a^2} \quad \text{or} \quad H_\alpha = \frac{2J}{ca^2}r.$$

In the second domain,

$$H_\alpha \cdot 2\pi r = \frac{4\pi}{c}J \quad \text{or} \quad H_\alpha = \frac{2J}{cr}.$$

In the third domain, $I = 0$ and $H_\alpha = 0$.

Hence, we get the following spatial distribution of the magnetic field:

$$H_r = H_z = 0, \quad H_\alpha = \begin{cases} \dfrac{2Jr}{ca^2}, & \text{if } r < a; \\[2mm] \dfrac{2J}{cr}, & \text{if } a \le r \le b; \\[2mm] 0, & \text{if } r > b. \end{cases}$$

Example 4.5. Determine the magnetic field $\vec{H}$ which is created by two parallel planes, where the currents with the same surface density $i = \text{const}$ flow. Consider two cases: (a) currents flow in the opposite directions; (b) currents have the same direction.

Solution. The magnetic field of two planes with currents is equal to the vector sum of the magnetic fields created by each plane with current:

$$\vec{H} = \vec{H}_1 + \vec{H}_2.$$

It is obvious that, for opposite directions of the currents, the magnetic fields $\vec{H}_1$ and $\vec{H}_2$ will have the same direction in the domain between the planes and the opposite directions outside the planes. Since, by condition, the values of currents are identical, we have $|\vec{H}_1| = |\vec{H}_2|$. Hence, in case a): $H = 0$ outside the planes, and $H = 2H_1$ between the planes.

Let us find the magnetic field created by a plane with current.

If an electric current flows over a thin plate, we can neglect the plane thickness. It is convenient to characterize the current distribution over the surface by a linear density. The linear current density is an analog of the surface charge density. Hence, $i = \frac{dI}{dl}$, which yields $I = \oint idl$.

Let us choose a rectangular contour for the calculation of the line integral of the vector $\vec{H}_1$ so that the contour plane is perpendicular to the plane and the current direction, and two sides of the contour (let their lengths be l) are parallel to the plate. The line integral of the vector $\vec{H}_1$ over the chosen contour is

$$\oint H_{1l}dl = 2H_1l.$$

On the other hand,

$$\oint H_{1l}dl = \frac{4\pi}{c}I = \frac{4\pi}{c}il.$$

Finally, we have $H_1 = \frac{2\pi}{c}i$; $H = \frac{4\pi}{c}i$.

If the currents have the same direction, the resulting magnetic field will be nonzero in the domain outside the planes. Between the planes, the magnetic fields $\vec{H}_1$ and $\vec{H}_2$ are directed in opposite directions. Hence, their sum is zero.

Thus: (a) between the planes, $H = \frac{4\pi}{c}i$, and $H = 0$ outside them; (b) between the planes, $H = 0$, and $H = \frac{4\pi}{c}i$ outside them. In both cases, the magnetic field is directed perpendicularly to the current and in parallel to the planes.

Example 4.6. Two thin cylindrical shells with radii a and b $(a < b)$ are coaxial, and the space between them is filled with a substance with magnetic permeability μ. Find the coefficient of self-induction per unit length L.

Solution. According to formula (4.13) and with regard for relation (4.8), the energy of the magnetic field per unit length of the infinitely long direct wire is given by the integral

$$W = \frac{\mu}{8\pi} \int H^2 \, 2\pi r dr.$$

The magnetic field of a cylindrical conductor $H = \frac{2I}{cr}$. Substituting it in the formula for the energy and executing the integration with respect to r from a to b, we get

$$W = \frac{\mu I^2}{c^2} \ln \frac{b}{a}.$$

Comparing relation (4.17) with formula (4.14), we get the formula for the coefficient of self-induction per unit length:

$$L = 2\mu \cdot \ln \frac{b}{a}.$$

Problems

4.12. Find the strength of the magnetic field inside and outside a cylindrical conductor in which a current uniformly distributed over its cross-section flows with density j. The conductor radius is R.

Answer.

$$\vec{H} = \frac{2\pi}{c}[\vec{j} \times \vec{r}], \quad r < R; \quad \vec{H} = \frac{2\pi R^2}{cr^2}[\vec{j} \times \vec{r}], \quad r > R,$$

where r is the distance from the cylinder axis.

4.13. Find the strength of the magnetic field inside a cylindrical cavity in a cylindrical conductor, in which a current uniformly distributed over its cross-section flows with density j. The axes of the cylindrical cavity and the cylindrical conductor are parallel and are placed at the distance a from each other.

Answer. $\vec{H} = \frac{2\pi}{c}[\vec{j} \times \vec{a}]$.

4.14. Find the strength of the magnetic field of a plane, where a current flows with surface density i, which is the same at any point of the plane.

Answer.

$$H_y = \begin{cases} -\dfrac{2\pi}{c}i, & x < 0, \\[2mm] \dfrac{2\pi}{c}i, & x < 0. \end{cases}$$

4.15. A current is uniformly distributed over an infinite rectilinear strip of width a. The surface current density is i. Find the strength of the magnetic field. Consider a special case where the strip width tends to infinity. Compare with the solution of Problem 4.14.

Answer.

$$H_x = \frac{i}{c}\ln\frac{x^2 + (y - a/2)^2}{x^2 + (y + a/2)^2},$$

$$H_y = \frac{2i}{c}\left(-\arctan\frac{y - a/2}{x} + \arctan\frac{y + a/2}{x}\right),$$

$$H_z = 0.$$

As $a \to \infty$, the answer coincides with the solution of Problem 4.14.

4.16. Along two infinite linear conductors placed at the distance d from each other, two currents I flow in opposite directions. Calculate the vector potential of the system.

Answer. $A_z = \frac{2}{c}I\ln\frac{r_2}{r_1}$, where r_1 and r_2 are distances from the observation point to the first and second conductors, respectively.

4.17. Find the vector potential and the strength of the magnetic field created by a current I flowing over a ring with radius R. Consider the case where the observation point is located on the ring axis.

Answer.

$$A_\varphi = \frac{4\mu_0 I}{ck}\left(\frac{R}{\rho}\right)^{1/2}\left[\left(1 - \frac{1}{2}k^2\right)K(k) - E(k)\right],$$

where ρ is the distance from the observation point to the ring axis, $k^2 = \frac{4R\rho}{(R+\rho)^2+z^2}$; $K(k) = \int_0^{\pi/2}(1 - k^2\sin^2\theta)^{-1/2}d\theta$ is the elliptic integral of the first kind; $E(k) = \int_0^{\pi/2}(1 - k^2\sin^2\theta)^{1/2}d\theta$ is the elliptic integral of the second kind,

$$H_\rho = \frac{2I}{c}\frac{z}{\rho[(R+\rho)^2 + z^2]^{1/2}} \cdot \left[-K(k) + \frac{R^2 + \rho^2 + z^2}{(R - \rho)^2 + z^2}E(k)\right];$$

$$H_z = \frac{2I}{c} \frac{1}{\rho[(R+\rho)^2 + z^2]^{1/2}} \cdot \left[K(k) + \frac{R^2 - \rho^2 - z^2}{(R-\rho)^2 + z^2} E(k) \right];$$

$$H_\varphi = 0.$$

On the symmetry axis, $\rho \to 0$, $H_\rho \to 0$, $H_z = \frac{2\pi}{c} \frac{R^2 I}{(R^2 + z^2)^{1/2}}$.

4.18. A cylinder with radius R_2 contains a conductor with radius R_1, whose magnetic permeability is μ_1. Let the medium with magnetic permeability μ_2 be placed between the conductor and the cylinder. Determine the inductance L of a unit length of the contour.

Answer.

$$L = \frac{\mu_1}{2} + 2\mu_2 \ln \frac{R_2}{R_1}.$$

4.19. Show that a constant homogeneous magnetic field, whose induction is $\vec{B}$, can be described by the vector potential $\vec{A} = \frac{1}{2}[\vec{B} \times \vec{r}]$.

4.20. Find the eigenfrequencies of two inductively coupled contours with coefficients of self-induction L_1 and L_2, coefficient of mutual induction L_{12}, capacities C_1 and C_2, and zero active resistances.

Answer.

$$\omega_{12}^2 = \frac{L_1 C_1 + L_2 C_2 \mp [(L_1 C_1 - L_2 C_2)^2 + 4 C_1 C_2 L_{12}^2]^{1/2}}{2 C_1 C_2 (L_1 L_2 - L_{12}^2)}.$$

Section 5

Electromagnetic Waves

5.1. Wave equation and its simple solutions

The Maxwell equations are as follows:

$$\text{rot}\,\vec{E}(\vec{r}, t) = -\frac{1}{c}\frac{\partial \vec{H}(\vec{r}, t)}{\partial t}; \tag{5.1}$$

$$\text{rot}\,\vec{H}(\vec{r},\ t) = \frac{1}{c}\frac{\partial \vec{E}(\vec{r},\ t)}{\partial t} + \frac{4\pi}{c}\vec{j}(\vec{r},\ t); \tag{5.2}$$

$$\text{div}\,\vec{D}(\vec{r}, t) = 4\pi\rho(\vec{r}, t); \tag{5.3}$$

$$\text{div}\,\vec{B}(\vec{r}, t) = 0. \tag{5.4}$$

where $\vec{D} = \varepsilon_0\vec{E},\ \vec{B} = \mu_0\vec{H}$.

The energy density of the electromagnetic field

$$w = \frac{1}{8\pi}(E^2 + H^2),$$

and Poynting's vector is

$$\vec{\gamma} = \frac{c}{4\pi}\vec{E} \times \vec{H}.$$

For the electromagnetic field in the case where $\rho_0 = 0$ and $j_0 = 0$, we have

$$\text{rot}\,\vec{E}(\vec{r}, t) = -\frac{1}{c}\frac{\partial \vec{H}(\vec{r}, t)}{\partial t};$$

$$\text{rot}\,\vec{H}(\vec{r},\ t) = \frac{1}{c}\frac{\partial \vec{E}(\vec{r},\ t)}{\partial t};$$

$$\text{div}\, \vec{D}(\vec{r}, t) = 0;$$

$$\text{div}\, \vec{B}(\vec{r}, t) = 0.$$

The partial polarization of waves:

— tensor of polarization

$$J_{\alpha\beta} = \overline{E_\alpha(t) E_\beta^*(t)}, \quad \alpha, \beta = 1, 2; \tag{5.5}$$

— field strength:

$$J_{\alpha\alpha} = I.$$

Lorentz gauge:

$$\text{div}\vec{A} + \frac{1}{c}\frac{\partial \varphi}{\partial t} = 0.$$

Coulomb gauge:

$$\text{div}\vec{A} = 0.$$

The wave equation or the d'Alembert homogeneous equation reads

$$\Delta \vec{E} - \frac{1}{c^2}\frac{\partial^2 \vec{E}}{\partial t^2} = 0,$$

$$\Delta \vec{H} - \frac{1}{c^2}\frac{\partial^2 \vec{H}}{\partial t^2} = 0.$$

Example 5.1. Write the d'Alembert equation and the Lorentz condition for Fourier components of the potentials $\varphi(\vec{r}, t)$ and $\vec{A}(\vec{r}, t)$. Consider all three versions of Fourier expansions.

Solution. Expansion in harmonic components:

$$\Delta \vec{A}_\omega + \frac{\omega^2}{c^2}\vec{A}_\omega = -\frac{4\pi}{c}\vec{j}_\omega;$$

$$\Delta \varphi_\omega + \frac{\omega^2}{c^2}\varphi_\omega = -4\pi\rho_\omega;$$

$$\text{div}\vec{A}_\omega - \frac{i\omega}{c}\varphi_\omega = 0.$$

Expansion in plane waves:

$$\ddot{\vec{A}}_{\vec{k}} + k^2 c^2 \vec{A}_{\vec{k}} = 4\pi c \vec{j}_{\vec{k}};$$

$$\ddot{\varphi}_{\vec{k}} + k^2 c^2 \varphi_{\vec{k}} = 4\pi c^2 \rho_{\vec{k}};$$

$$i c \vec{k} \cdot \dot{\vec{A}}_{\vec{k}} + \dot{\varphi}_{\vec{k}} = 0.$$

Expansion in plane monochromatic waves:

$$\left(k^2 - \frac{\omega^2}{c^2} \right) \vec{A}_{\vec{k}\omega} = \frac{4\pi}{c} \vec{j}_{\vec{k}\omega};$$

$$\left(k^2 - \frac{\omega^2}{c^2} \right) \varphi_{\vec{k}\omega} = 4\pi \rho_{\vec{k}\omega};$$

$$\vec{k} \cdot \vec{A}_{\vec{k}\omega} - \frac{\omega}{c} \varphi_{\vec{k}\omega} = 0.$$

Example 5.2. In the general case of complex amplitude $\vec{E}_0 = \vec{E}_{01} + i\vec{E}_{02}$, where $\vec{E}_{01}, \vec{E}_{02}$ are real vectors, we can separate two mutually perpendicular real vectors $\vec{\varepsilon}_{01}, \vec{\varepsilon}_{02}$ such that

$$\vec{E}_0 = (\vec{\varepsilon}_{01} + i\vec{\varepsilon}_{02})e^{i\alpha}, \quad \vec{E}_1 \cdot \vec{E}_2 = 0,$$

and $\alpha(-\pi < \alpha \le \pi)$ is some initial phase.

(a) Express the initial phase in terms of the vectors $\vec{E}_{01}, \vec{E}_{02}$.
(b) Show that the observed field (real part of the complex vector $\vec{E}$) can be written in the form

$$\vec{E} = \vec{\varepsilon}_1 \cos(\vec{k} \cdot \vec{r} - \omega t + \alpha) - \vec{\varepsilon}_2 \sin(\vec{k} \cdot \vec{r} - \omega t + \alpha).$$

(c) Show that the end of a vector $\vec{E}$ traces an ellipse (*elliptic polarization*), or a circle (*circular polarization*) at a certain point in space or oscillates around some straight line (*linear polarization*). In the case of elliptic or circular polarization, two opposite directions of rotation are possible. For linear polarization, the oscillations can occur in two mutually perpendicular directions. Therefore, for a given direction of propagation of a wave, there are two different independent types of polarization.
(d) How can the direction of rotation of the vector $\vec{E}$ relative to the direction of propagation of the wave be determined?

Solution. In order to define the directions of rotation, we write Eq. (5.5) for projections on the coordinate axes, by choosing the right-handed coordinate system, which is usually in use, with the axis Ox along $\vec{\varepsilon}_1$ and the axis Oz in the direction of propagation of the wave $\vec{k}$. We write the argument of trigonometric functions so that it increases with t. We get

$$E_x = E_1 \cos(\omega t - \vec{k} \cdot \vec{r} - \alpha);$$

$$E_y = \pm E_2 \sin(\omega t - \vec{k} \cdot \vec{r} - \alpha),$$

where $E_1 \geq 0$ and $E_2 \geq 0$. The signs "+" and "−" in the second formula correspond to the right and left triples of the vectors $\vec{\varepsilon}_1$, $\vec{\varepsilon}_2$, $\vec{k}$.

For the sign "+", the wave has right helicity, i.e., the direction of rotation of the vector $\vec{E}$ and the direction of propagation form a right screw. For the sign "−", the helicity is left (a screw with left thread). Due to some historical reasons, researchers use the opposite terminology in optics: the rotation of the vector $\vec{E}$ in the direction of right screw is called *left*, and the opposite rotation is called *right*.

$$\tan 2\alpha = 2\vec{E}_{01} \cdot \vec{E}_{02}/(E_{01}^2 + E_{02}^2).$$

Example 5.3. A plane monochromatic wave with the intensity I propagates along the axis Oz and is polarized over an ellipse with semiaxes a and b. The major semiaxis a forms an angle ϑ with the axis Ox. Construct the tensor of polarization and consider possible special cases.

Solution. Let us introduce the rectangular axes $x' \| a$ and $y' \| b$. In these axes, the complex amplitude of the field takes the form

$$\vec{E}_0 = a\vec{e}_{x'} \pm ib\vec{e}_{y'},$$

where the signs "+" and "−" correspond to the right and left elliptic polarizations.

The intensity $I = a^2 + b^2$. For the x'-component of the field, we set the zero phase. By representing the unit vectors $\vec{e}_{x'}$, $\vec{e}_{y'}$ in terms of $\vec{e}_x$, $\vec{e}_y$, we get

$$I_{11} = a^2 \cos^2 \vartheta + b^2 \sin^2 \vartheta;$$

$$I_{22} = a^2 \sin^2 \vartheta + b^2 \cos^2 \vartheta;$$

$$I_{12} = (a^2 - b^2) \sin \vartheta \cos \vartheta \pm 2iab = I_{21}^*.$$

The upper and lower signs correspond, respectively, to the right and left elliptic polarizations. If $b = 0$, the polarization is linear, and the tensor I_{ik} takes the form

$$I_{ik} = I \begin{pmatrix} \cos^2 \vartheta & \sin \vartheta \cos \vartheta \\ \sin \vartheta \cos \vartheta & \sin^2 \vartheta \end{pmatrix}.$$

If $a = b = \sqrt{I/2}$, the polarization is circular, and

$$I_{ik} = \frac{I}{2} \begin{pmatrix} 1 & \mp i \\ \pm i & 1 \end{pmatrix}.$$

Example 5.4. The position of some object can be determined with radar. What is the limiting accuracy of measurements, if the distance to the object is l, and the wavelength is λ?

Solution. The wave pulse sent by a radar has a width Δx, which is connected with the transverse scattering of wave vectors $k_\perp$ by the relation $\Delta x \, k_\perp \geq 1$. On the other hand, we have, obviously, $\Delta x / l \approx k_\perp / k$. From these two relations, we get the uncertainty of the determination of a position of the object:

$$\Delta x \geq \sqrt{l\lambda}.$$

Example 5.5. Construct the Hamilton function of a free electromagnetic field in the principal coordinates and write the field equations in the Hamilton form.

Solution. Let us calculate the field energy

$$W = \frac{1}{8\pi} \int_V (E^2 + H^2) dV \qquad (5.7)$$

in the basic domain V. For this purpose, we define the field strengths

$$\vec{E} = -\frac{1}{c}\dot{\vec{A}} = -\frac{1}{c}\sum_{\vec{k}\sigma}\dot{q}_{\vec{k}\sigma}\vec{A}_{\vec{k}\sigma} = -\frac{1}{c}\sum_{\vec{k}\sigma}\dot{q}^*_{\vec{k}\sigma}\vec{A}^*_{\vec{k}\sigma};$$

$$\vec{H} = \mathrm{rot}\vec{A} = i\sum_{\vec{k}\sigma}q_{\vec{k}\sigma}[\vec{k}\times\vec{A}_{\vec{k}\sigma}] = -i\sum_{\vec{k}\sigma}q^*_{\vec{k}\sigma}[\vec{k}\times\vec{A}^*_{\vec{k}\sigma}].$$

In view of the condition of orthogonality, Eq. (5.7) yields

$$W = \frac{(A^0)^2 V}{8\pi c^2}\sum_{\vec{k}\sigma}(\dot{q}^*_{\vec{k}\sigma}(t)\dot{q}_{\vec{k}\sigma}(t) + c^2 k^2 q^*_{\vec{k}\sigma}(t)q_{\vec{k}\sigma}(t)).$$

Substituting $q_{\vec{k}\sigma}(t) = b_{\vec{k}\sigma}(t) + b^*_{-\vec{k}\sigma}(t)$, we get

$$W = \frac{(A^0)^2 V}{2\pi c^2}\sum_{\vec{k}\sigma}\omega_k^2 b^*_{\vec{k}\sigma}(t)b_{\vec{k}\sigma}(t). \tag{5.8}$$

We have obtained the total energy of the field in a volume V in the form of a sum of the energies of separate eigenoscillations presented in terms of the complex principal coordinates. We now introduce the real variables $Q_{\vec{k}\sigma}$, $P_{\vec{k}\sigma}$:

$$b_{\vec{k}\sigma} = \frac{1}{2}\left(Q_{\vec{k}\sigma} + \frac{iP_{\vec{k}\sigma}}{\omega_k}\right). \tag{5.9}$$

It follows from their definition (5.9) that these variables also satisfy the equation of motion of a harmonic oscillator. The energy of the system presented in terms of the generalized coordinates and momenta is called the Hamilton function of the system.

We denote the Hamilton function by $\mathcal{H}$. Substituting the complex coordinates written in terms of the real variables (5.9) in Eq. (5.8), we obtain

$$\mathcal{H} = \frac{(A^0)^2 V}{8\pi}\sum_{\vec{k}\sigma}(P_{\vec{k}\sigma}^2 + \omega_k^2 Q_{\vec{k}\sigma}^2).$$

At this stage, it is convenient to choose the normalization constant A^0 so that the effective masses of oscillators of the field become equal to 1:

$$A^0 = \sqrt{\frac{4\pi c^2}{V}}.$$

Finally, the Hamilton function of the field takes the canonical form

$$\mathcal{H} = \sum_{\vec{k}\sigma} \mathcal{H}_{\vec{k}\sigma}, \quad \text{where } \mathcal{H}_{\vec{k}\sigma} = \frac{1}{2}(P_{\vec{k}\sigma}^2 + \omega_k^2 Q_{\vec{k}\sigma}^2). \tag{5.10}$$

The canonical equations or the Hamilton equations follow from the Hamilton functions (5.10) in the customary way,

$$\dot{Q}_{\vec{k}\sigma} = \frac{\partial \mathcal{H}}{\partial P_{\vec{k}\sigma}} = P_{\vec{k}\sigma}, \quad \dot{P}_{\vec{k}\sigma} = -\frac{\partial \mathcal{H}}{\partial Q_{\vec{k}\sigma}} = -\omega_k^2 Q_{\vec{k}\sigma},$$

and yield the proper equation of motion

$$\ddot{Q}_{\vec{k}\sigma} + \omega_k^2 Q_{\vec{k}\sigma} = 0.$$

This proves that Q, and P are the canonical variables for oscillators of the field.

Example 5.6. Two plane monochromatic linearly polarized waves emitted by a single particle propagate along the axis Oz. The first wave is polarized along Ox and has amplitude a. The second wave is polarized along Oy, has amplitude b, and is ahead of the first one by the phase χ. Study the polarization of the resulting wave as a function of a/b.

Solution. By applying the method of Example 5.2, we can write

$$\vec{E}_0 = \vec{E}_1 + \vec{E}_2 = a\vec{e}_x + be^{i\chi}\vec{e}_y = (\varepsilon_1 + i\varepsilon_2)e^{i\alpha}$$

and get

$$\vec{E}_1 = a\cos\alpha\,\vec{e}_x + b\cos(\chi - \alpha)\vec{e}_y, \tag{5.11}$$

$$\vec{E}_2 = -a\sin\alpha\,\vec{e}_x + b\sin(\chi - \alpha)\vec{e}_y, \tag{5.12}$$

$$\tan 2\alpha = \frac{b^2 \sin 2\chi}{a^2 + b^2 \cos 2\chi}, \quad -\pi < \alpha \leq \pi. \tag{5.13}$$

Note that Eq. (5.11) is obtained under the condition $\vec{E}_1 \cdot \vec{E}_2 = 0$. The helicity of the resulting wave is defined by the sign of the product $\vec{E}_2 \cdot \vec{e}_{y'}$, where $\vec{e}_{y'} = \vec{n} \times \vec{E}_1/E_1$ is the third unit vector that composes the right triple with $\vec{e}_{x'} = \vec{E}_1/E_1$ and $\vec{n}$. Using formulas (5.11)–(5.13), we find $\vec{E}_2 \cdot \vec{e}_{y'} = ab\sin\chi/E_1$.

According to the results of Example 5.2, if $ab > 0$ and $\sin\chi > 0$, $0 < \chi < \pi$, the helicity is right. If $\sin\chi < 0$, $-\pi < \chi < 0$, it is left. If $\chi = 0$ and $\chi = \pm\pi$, then the polarizations are linear in two mutually perpendicular planes.

Problems

5.1. A ball with radius a and charge e uniformly distributed over the volume rotates with a variable angular velocity $\Omega(t)$ around one of its axes. Write the densities $\rho(\vec{r}, t), \vec{j}(\vec{r}, t)$, and check if the equation of continuity $\frac{\partial\rho}{\partial t} + \mathrm{div}\vec{j} = 0$ is satisfied.

Answer. $\rho = \frac{3e}{4\pi a^3}\Theta(a - r)$; $\vec{j} = \rho\vec{\Omega} \times \vec{r}$, where Θ is a step-function.

5.2. Let an electromagnetic perturbation with finite duration pass through a unit area. Express the spectral density of the energy flow of the electromagnetic field Γ_ω in terms of the Fourier harmonics $\vec{E}_\omega$ and $\vec{H}_\omega$. The quantity Γ_ω is normalized by the condition

$$\int_0^\infty \Gamma_\omega d\omega = \Gamma = \frac{c}{4\pi}\int_{-\infty}^\infty \vec{E}(t) \times \vec{H}(t)\,dt,$$

where Γ is the total density of the energy flow through a unit area for the whole perturbation passage duration.

Answer.

$$\Gamma_\omega = \frac{c}{4\pi^2}\mathrm{Re}[\vec{E}_\omega \times \vec{H}_\omega^*].$$

5.3. Using the Maxwell equations (5.1)–(5.4), show that the plane waves satisfy the *conditions of transversality*

$$\vec{n} \cdot \vec{E} = 0, \quad \vec{n} \cdot \vec{H} = 0.$$

For plane waves propagating in the same direction, show that the electric and magnetic vectors are mutually perpendicular and are connected by the relations

$$\vec{H} = \vec{n} \times \vec{E}, \quad \vec{E} = \vec{H} \times \vec{n}, \quad \vec{E} = \vec{H}.$$

Find the connection between the electromagnetic energy density and the energy flow density (Poynting's vector) for such waves.

Answer. $\vec{\gamma} = cw\vec{n}$, $w = (E^2 + H^2)/8\pi = E^2/4\pi$. The energy is transferred in the space with velocity c.

5.4. An electromagnetic wave is a superposition of two noncoherent "almost monochromatic" waves with the same intensity I with approximately equal frequencies and wave vectors. Both waves are polarized linearly, and the directions of polarization are set in the plane perpendicular to their wave vector by the unit vectors $\vec{e}^{(1)}(1,0)$ and $e^{(2)}(\cos\vartheta, \sin\vartheta)$. Construct the tensor of polarization I_{ik} of the resulting partially polarized wave and determine the degree of its polarization. Clarify the character of the polarization of this wave.

Answer. The amplitude of the summary wave

$$\vec{E} = \vec{E}_1 + \vec{E}_2 = E(\vec{e}^{(1)} + \vec{e}^{(2)}e^{i\alpha}),$$

where α is the randomly varying phase shift, and $|\vec{E}|^2 = I$.

By definition, the components of the tensor of polarization are

$$I_{ik} = \overline{E_i E_k^*} = \overline{I(\vec{e}^{(1)} + \vec{e}^{(2)}e^{i\alpha})}_i(\vec{e}^{(1)} + \vec{e}^{(2)}e^{-i\alpha})_k.$$

By averaging over the time, we obtain $\overline{e^{\pm i\alpha}} = 0$, and the tensor of polarization takes the form

$$I_{ik} = I\begin{pmatrix} 1 + \cos^2\vartheta & \sin\vartheta\cos\vartheta \\ \sin\vartheta\cos\vartheta & 1 - \cos^2\vartheta \end{pmatrix}.$$

This yields

$$P = |\cos\vartheta|.$$

The same result can be obtained if we diagonalize the tensor I_{ik}. Its principal values are $I_1 = 1 + |\cos \vartheta|$, $I_2 = 1 - |\cos \vartheta|$. From this, we have again $P = (I_1 - I_2)/(I_1 + I_2) = |\cos \vartheta|$. The basis vectors are $\vec{e}_1 = (\cos \frac{\vartheta}{2}, \sin \frac{\vartheta}{2})$ and $\vec{e}_2 = (-\sin \frac{\vartheta}{2}, \cos \frac{\vartheta}{2})$. In this case, they are real.

The resulting wave is composed of the nonpolarized part with the intensity $I(1 - |\cos \vartheta|)$ and the part linearly polarized along the direction $\vec{e}_1 = (\cos \frac{\vartheta}{2}, \sin \frac{\vartheta}{2})$ with the intensity $I|\cos \vartheta|$:

$$(I_{ik}) = I(1 - |\cos \vartheta|)(\delta_{ik}) + I|\cos \vartheta|$$

$$\times \begin{pmatrix} \cos^2 \dfrac{\vartheta}{2} & \sin \dfrac{\vartheta}{2} \cos \dfrac{\vartheta}{2} \\ \sin \dfrac{\vartheta}{2} \cos \dfrac{\vartheta}{2} & \sin^2 \dfrac{\vartheta}{2} \end{pmatrix}.$$

The resulting wave is completely polarized (but it is not monochromatic), if $\vartheta = 0$. If $\vartheta = \pi/2$, we observe the full depolarization.

5.5. The wave package $\psi(x, t)$ is formed by a superposition of plane monochromatic waves with different frequencies in the free space. If $x = 0$, $\psi(0, t) = u(t)$, the package shape is known. Find the amplitude function $\psi(k)$.

Answer.

$$\psi(k) = \frac{1}{2\pi} \int_{-\infty}^{\infty} u(t) e^{ikct} dt.$$

5.6. In the previous problem, study the dependence of the polarization on the phase shift α in the case $a = b$.

Answer. If $\alpha = 0$, the polarization is linear, and the polarization plane passes through the bisectrix of the angle between the axes Ox, Oy. If $\alpha = \pi$, the polarization is also linear, and the polarization plane passes through the bisectrix of the angle between the axes Ox and $-Oy$. In the case $\alpha = \pi/2$ the polarization is circular with right helicity (Fig. 5.1, a).

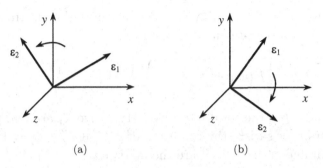

(a)   (b)

Fig. 5.1

If $\alpha = -\pi/2$, the polarization is circular with left helicity (Fig. 5.1, b). In the other cases, it is elliptic. The helicity is right or left under the conditions $0 < \alpha < \pi$ or $-\pi < \alpha < 0$.

5.7. Solve Problem 5.4 in the case where the intensities of the waves are different ($I_1 \neq I_2$), and the directions of polarization form an angle of $\pi/4$.

Answer. The tensor of polarization

$$I_{ik} = \begin{pmatrix} I_1 + I_2/2 & I_2/2 \\ I_2/2 & I_2/2 \end{pmatrix}$$

(axis x_1 coincides with the direction of polarization of the first wave).

The degree of polarization

$$P = \frac{2\sqrt{I_1^2 + I_2^2}}{I_1 + I_2 + \sqrt{I_1^2 + I_2^2}}.$$

The resulting wave is composed of the nonpolarized wave with intensity $(I_1 + I_2)(1 - P)/2$ and a linearly polarized wave. The direction of the linear polarization forms the angle

$$\vartheta = \text{arctg}\, \frac{2I_2\sqrt{I_1^2 + I_2^2}}{I_1(I_1 + I_2) + (3I_1 + 2I_2)\sqrt{I_1^2 + I_2^2}}$$

with the direction of polarization of the first wave.

5.8. The Hermitian tensor of polarization of the electromagnetic wave can be presented in the form

$$I_{ik} = \frac{1}{2}I\left(\delta_{ik} + \sum_{l=1}^{3} \zeta_l \hat{\tau}_{ik}^{(l)}\right) = \frac{1}{2}I\begin{pmatrix} 1 + \zeta_3 & \zeta_1 - i\zeta_2 \\ \zeta_1 + i\zeta_2 & 1 - \zeta_3 \end{pmatrix},$$

where I is the total intensity of the wave; ζ_i are real parameters that satisfy the condition $\zeta^2 = \zeta_1^2 + \zeta_2^2 + \zeta_3^2 \leq 1$ (Stokes parameters); and $\hat{\tau}^{(l)}$ are the matrices:

$$\hat{\tau}^{(1)} = \begin{pmatrix} 0 & 1 \\ 1 & 0 \end{pmatrix}, \quad \hat{\tau}^{(2)} = \begin{pmatrix} 0 & -i \\ i & 0 \end{pmatrix}, \quad \hat{\tau}^{(3)} = \begin{pmatrix} 1 & 0 \\ 0 & -1 \end{pmatrix}.$$

Clarify the physical meaning of the parameters ζ_i. For this purpose, express the degree of polarization ρ of the wave in terms of ζ_i and determine the polarizations of two basic waves, into which a partially polarized wave decays, in the following three cases:

$$\text{(a) } \zeta_1 \neq 0, \quad \zeta_2 = \zeta_3 = 0; \quad \text{(b) } \zeta_2 \neq 0,$$

$$\zeta_1 = \zeta_3 = 0; \quad \text{(c) } \zeta_3 \neq 0, \quad \zeta_1 = \zeta_2 = 0.$$

Answer. $\rho = \frac{1-\zeta}{1+\zeta}$; if $\zeta = 0$, the wave is not polarized, if $\zeta = 1$, it is completely polarized.

Assume that $\zeta_i = \zeta \eta_i$, where $\eta_1^2 + \eta_2^2 + \eta_3^2 = 1$. Then

$$I_{ik} = \frac{I}{2}(1 - \zeta)\delta_{ik} + \frac{I\zeta}{2}\left(1 + \sum_{l=1}^{3} \eta_l \tau_{ik}^{(l)}\right).$$

The first and second terms in this formula correspond, respectively, to the completely nonpolarized and completely polarized states. In case (a), $\eta_3 = 1$, $\eta_1 = \eta_2 = 0$.

Comparing

$$I_{ik} = I\zeta \begin{pmatrix} 1 & 0 \\ 0 & 0 \end{pmatrix}$$

with the expression $I_{ik} = I n_i n_k^*$, we see that, in this case, $n_1 = 1$, $n_2 = 0$. In other words, the tensor I_{ik}'' describes a wave polarized in the direction of the axis x (propagating in the direction z).

Analogously, it is easy to verify that, in case (b), $n_1 = 1$, $n_2 = n_3 = 0$, and the wave is linearly polarized in the direction inclined by $45°$ to the axis x. In case (c), $n_2 = 1$, $n_1 = n_3 = 0$, and the wave has a circular polarization.

5.9. Let the emission source create a decaying signal $U'(t) = A_0 \Theta(t) e^{-\gamma t/2} \sin \omega_0 t$, where $\Theta(t)$ is a step-function; γ is the decay constant, and $A_0 = \text{const}$. Under which conditions will the signal be quasimonochromatic? Find the energy distribution over frequencies, by using the notion of an analytic signal. Estimate the band of frequencies of this signal and the product $\Delta\omega\Delta t$.

Answer.

$$|a(\omega)|^2 = \frac{A_0^2}{(\omega - \omega_0)^2 + \gamma^2/4}.$$

The signal will have a narrow band (will be quasimonochromatic) under the condition $\gamma \ll \omega_0$. If Δt is defined as the time over which the signal intensity $I \propto |U(t)|^2$ decreases by e times, and if $\Delta\omega$ is defined as the scatter at which the spectral power decreases also by e times, then $\Delta t \Delta\omega = \sqrt{e - 1}/2$. The shape of the spectrum obtained in this problem is called *Lorentz contour*.

5.10. Let a circuit be composed of two coils joined in series, whose inductances are L_1 and L_2. The mutual induction of the coils is $\mathfrak{M}$. Find the total inductance of the circuit. How will the total inductance be changed if one coil is wound in the reverse direction?

Answer. $U = \frac{1}{2}\mathcal{L}_1 I_1^2 + \frac{1}{2}\mathcal{L}_2 I_2^2 - \mathfrak{M} I_1 I_2$, where I_1, I_2 are the currents in the coils; and $\mathfrak{M}$ is the coefficient of mutual induction.

By condition, $I_1 = I_2 = I$. Therefore, the electric energy of the circuit can be presented in the form $U = \frac{1}{2}\mathcal{L}I^2$, where $\mathcal{L} = \mathcal{L}_1 + \mathcal{L}_2 - 2\mathfrak{M}$. If the turns of one coil are rewound in the reverse direction, then the sign of $\mathfrak{M}$ is changed.

5.11. A plane electromagnetic wave with frequency ω is reflected from the mirror, which moves with velocity v in the direction of propagation of the wave. Using the Maxwell equations, find the frequency of the reflected wave, as measured by a static observer.

Answer.

$$\omega' = \omega\left[1 - \left(\frac{v}{c}\right)\right] / \left[1 + \left(\frac{v}{c}\right)\right].$$

5.12. Using the symmetry of the Maxwell equations and the form of the electric and magnetic fields of an oscillating electric dipole, find the field of an oscillating magnetic dipole. This field must coincide with the field of the dipole, which is formed by the contour with the current $i = i_0 \cos \omega t$ and with radius $a(a \ll c/\omega)$.

Answer.

$$B_\varphi = 0,$$

$$B_\theta = \frac{\pi a^2 i_0}{c^2} \sin \theta$$

$$\times \left[\left(-\frac{\omega^2}{c^2 r} + \frac{1}{r^3}\right) \cos \omega \left(t - \frac{r}{c}\right) - \frac{\omega}{cr^2} \sin \omega \left(t - \frac{r}{c}\right)\right];$$

$$B_r = \frac{2\pi a^2 i_0}{c^2} \cos \theta \left[\frac{1}{r^3} \cos \omega \left(t - \frac{r}{c}\right) - \frac{\omega}{cr^2} \sin \omega \left(t - \frac{r}{c}\right)\right];$$

$$\vec{E} = c[\vec{B} \times \vec{e}_r].$$

5.13. A plane-polarized wave falls normally on the surface of a non-magnetic medium with dielectric permeability ε and conductance σ^*. Find the coefficient of reflection R.

Answer.

$$R = \frac{(1 - n)^2 + \kappa^2}{(1 + n)^2 + \kappa^2},$$

where

$$n = \sqrt{\frac{\varepsilon}{2}} \left[\sqrt{1 + \left(\frac{\sigma^*}{\omega \varepsilon \varepsilon_0} \right)^2} + 1 \right]^{1/2};$$

$$\kappa = \sqrt{\frac{\varepsilon}{2}} \left[\sqrt{1 + \left(\frac{\sigma^*}{\omega \varepsilon \varepsilon_0} \right)^2} - 1 \right]^{1/2}.$$

5.14. Along the plane interface of two dielectrics with dielectric permeabilities ε_1 and $-|\varepsilon_2|$ that differ in sign, a surface wave, whose magnetic field strength is perpendicular to the direction of propagation of a (TM)-wave, propagates. Determine the dispersion law of such a wave.

Answer.

$$k^2 = \frac{\omega^2 \varepsilon_1 |\varepsilon_2|}{c^2 (|\varepsilon_2| - \varepsilon_1)}.$$

5.15. An electromagnetic wave falls on the plane surface of a semi-infinite crystal at the angle ϑ_1. The optical axis of the crystal is perpendicular to its surface. Determine the directions of propagation of the ordinary and extraordinary rays in the crystal. *Answer.* The equality of the projection of wave vectors on the interface implies that $\frac{\sin \vartheta_2'}{\sin \vartheta_1} = \sqrt{\varepsilon_\perp}$ for the ordinary wave and

$$k_1 \sin \vartheta_1 = k_2 \sin \vartheta_2'' = k_0 \sin \vartheta_2'' \left(\frac{\varepsilon_\perp \varepsilon_\parallel}{\varepsilon_\perp \sin^2 \vartheta_2'' + \varepsilon_\parallel \cos^2 \vartheta''} \right)^{1/2}$$

for the extraordinary wave; here, $k_0 = \omega_0/c$; ϑ_2'' is the angle between the wave vector in the crystal $\vec{k}_2$ and the optical axis. From this, we have

$$\tan \vartheta_2'' = \sqrt{\varepsilon_\parallel} \sin \vartheta_1 / \sqrt{\varepsilon_\perp (\varepsilon_\parallel - \sin^2 \vartheta_1)}.$$

The direction of propagation of the extraordinary ray θ'' is connected with the direction ϑ_2'' of the vector $\vec{k}_2$ by the relation $\tan \theta_2'' = \frac{\varepsilon_\perp}{\varepsilon_{\parallel 2}} \tan \vartheta''$. Therefore, $\tan \theta_2'' = \sqrt{\varepsilon_\perp} \sin \vartheta_1 / \sqrt{\varepsilon_\parallel (\varepsilon_\parallel - \sin^2 \vartheta_1)}$.

Section 6

Field of Moving Charges

6.1. Liénard–Wiechert potentials

The d'Alembert equation:

$$\left.\begin{array}{l} \Delta \vec{A} - \dfrac{1}{c^2}\dfrac{\partial^2 \vec{A}}{\partial t^2} = -\dfrac{4\pi}{c}\vec{j}(\vec{r},\, t), \\[3mm] \Delta \varphi - \dfrac{1}{c^2}\dfrac{\partial^2 \varphi}{\partial t^2} = -4\pi\rho(\vec{r},\, t). \end{array}\right\}$$

Lorentz condition:

$$\operatorname{div}\vec{A} + \frac{1}{c}\frac{\partial \varphi}{\partial t} = 0. \tag{6.1}$$

Green function of a wave package:

$$\Delta G - \frac{1}{c^2}\frac{\partial^2 G}{\partial t^2} = -4\pi\delta(\vec{r} - \vec{r}')\delta(t - t'), \tag{6.2}$$

$$\vec{A}(\vec{r}, t) = \frac{1}{c}\int G(\vec{r}, t; \vec{r}', t')j(\vec{r}', t')d^3r'dt'. \tag{6.3}$$

Retarded potentials:

$$\vec{A}(\vec{r}, t) = \frac{1}{c}\int \frac{\vec{j}(\vec{r}', t - |\vec{r} - \vec{r}'|/c)}{|\vec{r} - \vec{r}'|}dV',$$

$$\varphi(\vec{r}, t) = \int \frac{\rho(\vec{r}', t - |\vec{r} - \vec{r}'|/c)}{|\vec{r} - \vec{r}'|}dV'. \tag{6.4}$$

The time dependence of the distributions of charges and currents indicates that the field at a point r at the time moment t is determined by values of the quantities $\vec{j}$ and ρ at the point $\vec{r}'$ at the previous time moment $t' = t - R/c$, where $R = \vec{r} - \vec{r}'$ is the distance. The electromagnetic perturbations propagate in vacuum with velocity c.

The energy of emission in a solid angle unit can be written in terms of Poynting's vector:

$$\frac{dI}{d\Omega} = \gamma \cdot \vec{n} r^2 = \frac{cr^2}{4\pi} \vec{H}^2(\vec{r},\, t). \qquad (6.5)$$

The intensity over all directions of the emission after the integration over the whole solid angle is

$$I(r,\, t) = \frac{cr^2}{4\pi} \int \vec{H}^2(\vec{r},t) d\Omega = \frac{cr^2}{4\pi} \int \vec{E}^2(\vec{r},t) d\Omega.$$

Example 6.1. Write the equations for the irrotational (potential) and solenoidal parts of the electromagnetic field vectors $\vec{E}(\vec{r},t)$, $\vec{H}(\vec{r},t)$. Show that the potential part of the electric field describes the instant (unretarded) Coulomb field that is created by the distribution of charges at the same moment in time at which $\vec{E}^{\parallel}(\vec{r},\, t)$ is determined.

Solution. By the definition of the solenoidal and potential quantities, we have:

$$\nabla \cdot \vec{H}^{\perp} = \nabla \cdot \vec{E}^{\perp} = \nabla \cdot \vec{j}^{\perp} = 0;$$

$$\nabla \times \vec{H}^{\parallel} = \nabla \times \vec{E}^{\parallel} = \nabla \times \vec{j}^{\parallel} = 0.$$

According to the Maxwell equation $\nabla \cdot \vec{H} = 0$, the magnetic field is solenoidal, i.e., $\vec{H} = \vec{H}^{\perp}, \vec{H}^{\parallel} = 0$. The potential electric field satisfies the equations of electrostatics

$$\nabla \cdot \vec{E}^{\parallel}(\vec{r},\, t) = 4\pi\rho(\vec{r},\, t), \quad \nabla \times \vec{E}^{\parallel}(\vec{r},\, t) = 0,$$

where the time enters as a parameter. They describe the instant Coulomb interaction. The solenoidal components satisfy the equations

$$\nabla \times \vec{H}^{\perp} = \frac{1}{c}\frac{\partial \vec{E}^{\perp}}{\partial t} + \frac{4\pi}{c}\vec{j}^{\perp}, \quad \nabla \times \vec{E}^{\perp} = -\frac{1}{c}\frac{\partial \vec{H}^{\perp}}{\partial t}.$$

Example 6.2. The oscillations of two electric dipole oscillators have the same frequency ω, but they are shifted by $\frac{\pi}{2}$ in phase. The amplitudes of the dipole moments are equal to p and are directed at the angle φ to each other.

The distance between the oscillators is small as compared with the wavelength. Find the field $\vec{H}$ in the wave zone, angular distribution $\frac{dI}{d\Omega}$, and the total emission intensity $\bar{I}$.

Solution. Let the axis x be directed along the moment amplitude of the oscillator advanced in phase, and let the moments of both oscillators lie in the plane xy. By ϑ, α, we denote the polar angles of the unit vector $\vec{n}$, which indicates the direction of propagation of the wave.

We get

$$\vec{H}(\vec{r}, t) = \vec{H} e^{-i\omega t'} = \frac{\omega^2 p}{c^2 r} \{\vec{e}_\vartheta [\sin \alpha + i \sin(\alpha - \varphi)]$$

$$+ \vec{e}_\alpha [\cos \alpha + i \cos (\alpha - \varphi)] \cos \vartheta\}^{-i\omega t'};$$

$$\frac{dI}{d\Omega} = \frac{p^2 \omega^4}{8\pi c^3} \{2 - [\cos^2 \alpha + \cos^2 (\alpha - \varphi)] \sin^2 \vartheta\},$$

$$\bar{I} = \frac{2p^2 \omega^4}{3c^3}.$$

The emission is maximum in the directions $\vartheta = 0$ and $\vartheta = \pi$, which are perpendicular to the moments of both oscillators, and is nonuniformly distributed over azimuths. It is shown in Fig. 6.1 in the form of polar diagrams for $\varphi = 45°$.

Figure 6.1, a shows the angular distribution over α in the plane $\vartheta = 90°$. In Fig. 6.1, b, we present the angular distrbution over ϑ in the plane $\alpha = \frac{\varphi}{2} = 22.5°$.

Example 6.3. Find the electromagnetic field $\vec{H}, \vec{E}$ of a charge e which moves uniformly along a circle with radius a. The motion is relativistic, the angular velocity is ω. The distance to the observation point $r \gg a$. Find the time-averaged angular distribution $\overline{dI/d\Omega}$ and the total emission intensity $\bar{I}$ and study its polarization.

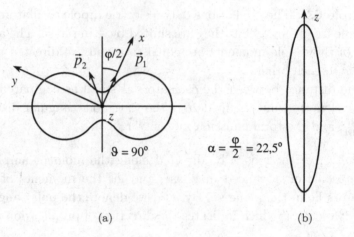

Fig. 6.1

Solution.

$$\vec{H} = \frac{1}{c}\frac{\partial \operatorname{rot} \vec{Z}}{\partial t} = ea\left[\vec{e}_\vartheta\left(-i\frac{\omega^2}{c^2 r} + \frac{\omega}{cr^2}\right)\right.$$

$$\left. + \vec{e}_\alpha\left(\frac{\omega^2}{c^2 r} + i\frac{\omega}{cr^2}\right)\cos\vartheta\right]e^{i(kr-\omega t+\alpha)};$$

$$\vec{E} = \operatorname{rot}\operatorname{rot}\vec{Z} = ea\left[\vec{e}_r\left(-\frac{i\omega}{cr^2} + \frac{1}{r^3}\right)2\sin\vartheta\right.$$

$$+ \vec{e}_\vartheta\left(\frac{\omega^2}{c^2 r} + i\frac{\omega}{cr^2} - \frac{1}{r^3}\right)\cos\vartheta$$

$$\left. + \vec{e}_\alpha\left(i\frac{\omega^2}{c^2 r} - \frac{\omega}{cr^2} - \frac{i}{r^3}\right)\right]e^{i(kr-\omega t+\alpha)}.$$

In the wave zone $\vec{r} \gg \lambda = \frac{2\pi c}{\omega}$, the formulas for $\vec{H}$, $\vec{E}$ are simplified:

$$\vec{H} = ea\frac{\omega^2}{c^2 r}(-i\vec{e}_\vartheta + \vec{e}_\alpha\cos\vartheta)e^{i(kr-\omega t+\alpha)};$$

$$\vec{E} = ea\frac{\omega^2}{c^2 r}(\vec{e}_\vartheta\cos\vartheta + i\vec{e}_\alpha)e^{i(kr-\omega t+\alpha)} = \vec{H}\times\vec{n}.$$

The emission in the upper hemisphere ($\cos \vartheta > 0$) has left elliptic polarization. In particular, if $\vartheta = 0$, we observe left circular polarization. As for the emission in the lower hemisphere ($\cos \vartheta < 0$), it has right elliptic polarization, and is circular for $\vartheta = \pi$. Waves emitted into the equatorial plane has a linear polarization. The angular distribution and the total emission intensity are as follows:

$$\frac{dI}{d\Omega} = \overline{\vec{\gamma}} \cdot \vec{n} r^2 = \frac{e^2 \omega^4 a^2}{8\pi c^3}(1 + \cos^2 \theta), \quad \bar{I} = \frac{2\omega^4 e^2 a^2}{3c^3}.$$

The case under study is realized, for example, during the motion of a charge in a homogeneous magnetic field.

Example 6.4. Study the state of the emission field polarization for a system of oscillators from Example 5.2, by using the same method.

Solution. By shifting the start of the reference phase by γ, we get a new amplitude of the field $\vec{H}e^{-i\gamma} = \vec{H}_1 - i\vec{H}_2$. Then we obtain

$$\tan 2\gamma = 2\frac{\sin \alpha \, \sin(\alpha - \varphi) + \cos \alpha \cos(\alpha - \varphi) \cos^2 \vartheta}{\sin^2 \alpha - \sin^2(\alpha - \varphi) + [\cos^2 \alpha - \cos^2(\alpha - \varphi)] \cos^2 \vartheta}.$$

From this formula, we can find $\cos \gamma$ and $\sin \gamma$ and determine $\vec{H}_1$ and $\vec{H}_2$ as functions of $\vartheta, \alpha, \varphi$.

Consider some special cases. If $\vartheta = 90°$, the polarization is linear; the polarization plane is perpendicular to the plane xy. If $\vartheta = 0, \pi$, the polarization is elliptic, and the ratio of semiaxes of the ellipse is $\tan \frac{\varphi}{2}$; in particular, if $\varphi = \frac{\pi}{2}$ and $\vartheta = 0, \pi$, the polarization is circular. The cases where $\alpha = \frac{\varphi}{2}, \frac{\varphi}{2} \pm \frac{\pi}{2}, \frac{\varphi}{2} + \pi$ can be easily studied as well. In all these cases, the polarization is elliptic. If $\alpha = \frac{\varphi}{2}, \frac{\varphi}{2} + \pi$, the polarization is circular in the directions that are determined by the conditions

$$\tan \frac{\varphi}{2} = |\cos \vartheta|,$$

If $\alpha = \frac{\varphi}{2} \pm \frac{\pi}{2}$, the directions with circular polarization are defined by the equation

$$\cot \frac{\varphi}{2} = |\cos \vartheta|.$$

Problems

6.1. Write the Fourier components $\vec{A}_\omega^A(\vec{r})$, $\varphi_\omega^A(\vec{r})$ of the advanced potentials, which are presented by formula (6.4) under the substitution of the advanced Green function G^A.

Answer.

$$\vec{A}_\omega^A(\vec{r}) = \frac{1}{c} \int \frac{\exp[-i\omega|\vec{r} - \vec{r}'|/c]}{|\vec{r} - \vec{r}'|} \vec{j}_\omega(\vec{r}') \, dV',$$

$$\varphi_\omega^A(\vec{r}) = \int \frac{\exp[-i\omega|\vec{r} - \vec{r}'|/c]}{|\vec{r} - \vec{r}'|} \rho_\omega(\vec{r}') \, dV'.$$

The Fourier harmonics of retarded potentials contain the exponential functions $\exp[i(kR - \omega t)]$, $k = \omega/c$, that corresponds to divergent spherical waves, which transfer a perturbation from the emitting system of charges in the environment. The Fourier harmonics of advanced potentials contain the exponential functions $\exp[-i(kR+\omega t)]$, which describe the spherical waves convergent to the center. Such waves would be created by some source located at infinity, rather than the given system of charges. Therefore, the advanced potentials are not suitable for the description of the process of emission in an infinite space.

6.2. Write the equations for the electromagnetic potentials $\varphi(\vec{r}, t)$ and $\vec{A}(\vec{r}, t)$, if they satisfy the condition $\text{div}\,\vec{A} = 0$ (Coulomb gauge) instead of the Lorentz condition (6.1). Express the equations in a form such that each equation includes one of the potentials and sources of the field.

Answer.

$$\Delta\varphi(\vec{r}, t) = -4\pi\rho(\vec{r},\, t),$$

$$\Delta\vec{A}(\vec{r}, t) - \frac{1}{c^2}\frac{\partial^2 \vec{A}}{\partial t^2} = -\frac{4\pi}{c}\vec{j}(\vec{r},\, t)$$

$$+ \frac{1}{c}\int\left\{\frac{\vec{j}(\vec{r}',\, t)}{R^3} - \frac{3\vec{R}(\vec{j}(\vec{r}',\, t)\cdot\vec{R})}{R^5}\right\}dV',$$

where $\vec{R} = \vec{r} - \vec{r}'$.

In the equation for φ, the time enters as a parameter. Therefore, the scalar potential describes a Coulomb field that is determined by the instant (nonretarded) distribution of charges.

6.3. A particle with charge e oscillates along the axis Oz by the law $z(t) = a \sin \omega t$. Calculate the emission intensities $dI_m/d\Omega$ at the multiple frequencies $\omega_m = m\omega$, $m = 1, 2, \ldots$, not assuming the ratio a/λ to be small.

Answer.

$$\vec{H}_m = i\frac{em\omega^2}{2\pi c^2} \frac{e^{ik_m r}}{r} \int_0^T \vec{n} \times \vec{v}(\tau) \exp i(m\omega\tau - \vec{k}_m \cdot \vec{s}(\tau))\, d\tau,$$

where $k_m = m\omega\vec{n}/c$, $\vec{s}(\tau) = \vec{e}_z z(\tau)$.

The integral over the time can be expressed in terms of the Bessel function. Finally, we get

$$\frac{dI_m}{d\Omega} = \frac{c}{2\pi}|\vec{H}_m|^2 r^2 = \frac{e^2\omega^2}{2\pi c}\tan^2\theta\, m^2 J_m^2(m\beta\,\cos\,\theta),$$

where $\beta = a\omega/c$, θ is the angle between the direction of emission $\vec{n}$ and the axis Oz. Here, we took into account that the harmonics with numbers m and $-m$ make an identical contribution to the emission. If $\beta \ll 1$, then

$$\frac{dI_1}{d\Omega} \approx \frac{e^2 a^2 \omega^4}{8\pi c^3}\sin^2\theta$$

• corresponds to the dipole emission;

$$\frac{dI_2}{d\Omega} \approx \frac{e^2 a^4 \omega^6}{8\pi c^5}\sin^2\theta\cos^2\theta$$

• corresponds to the quadrupole emission, which contains a small factor $(a\omega/c)^2$ as compared with the dipole one.

6.4. The simplest model of the emission of neutron stars (pulsars) is the model of oblique rotator: the ball with magnetic moment $\vec{m}$ rotates in vacuum with angular velocity ω around the axis that forms the angle φ with the direction of $\vec{m}$.

1. Calculate the time-averaged angular distribution $\overline{dI/d\Omega}$ and total emission intensity $\bar{I}$.

2. Estimate numerically the magnetic moment of a pulsar to the order of magnitude, taking the characteristic value of the magnetic field on the surfaces of neutron stars $H_0 \approx 2 \cdot 10^{12}$ Oe from observations and the theoretical value of the star radius $R \approx 10$ km.

3. Estimate numerically the emission intensity of a pulsar $\bar{I}$ and to compare it with Sun's luminosity $L_\odot \approx 4 \cdot 10^{33}$ erg/s, taking the period of rotation $T \approx 0.033$ s from the observations of the pulsar in the Crab-like nebula.

4. Compare the obtained intensity emission of the pulsar with the rate of decrease of the rotation energy of the star, which should be compared with the observational data on the increase of the period of rotation of the pulsar in the Crab-like nebula $\dot{T}/T \approx 1.3 \cdot 10^{-11}$ s^{-1}.

Answer. 1. We calculate the field strength in the wave zone by formulas (6.4), by taking $\vec{Q} = 1$. The angular distribution of the emission reads

$$\frac{\overline{dI}}{d\Omega} = \frac{1}{4\pi c^3}\overline{|\vec{n} \times \ddot{\vec{m}}|^2}. \tag{6.6}$$

To calculate the right-hand side, we use the equation of motion of the magnetic moment $\dot{\vec{m}} = \omega \times \vec{m}$. We get

$$(\vec{n} \times \ddot{\vec{m}})^2 = \omega^4 \vec{m}_\perp^2 (1 - \sin^2 \vartheta \cos^2(\omega t - \alpha)),$$

where $\vec{m}_\perp$ is the component perpendicular to the axis of rotation; ϑ is a polar angle counted from the direction of ω; ωt and α are asimuths of the vectors $\vec{m}_\perp$ and $\vec{n}$ in a plane perpendicular to ω. After the substitution in Eq. (6.6) and the averaging over the time, we obtain

$$\frac{\overline{dI}}{d\Omega} = \frac{\omega^4 m^2 \sin^2 \varphi}{8\pi c^3}(1 + \cos^2 \vartheta), \quad \bar{I} = \frac{2\omega^4 m^2 \sin^2 \varphi}{3c^3}. \tag{6.7}$$

2. Assuming the magnetic field of a pulsar to be a dipole one, we find $m \approx H_0 R^3 \approx 2 \cdot 10^{30}$ Oe·cm^3 to the order of magnitude.

3. Substituting the required quantity $\sin^2 \varphi \approx 1$ in Eq. (6.7), we get $\bar{I} \approx 1.3 \cdot 10^{38}$ erg/s, which is as high as $\approx 3 \cdot 10^4 L_\odot$.

4. The decrease in rotation energy can be calculated by the formula $\dot{\varepsilon}_{\text{rot}} = I\omega\dot{\omega} = -2\varepsilon_{\text{rot}}\dot{T}/T$, where $I = (2/5)MR^2$ is the moment of inertia of a ball; and $\varepsilon_{\text{rot}} \approx 2.6 \cdot 10^{33}$g is the star mass (Sun's mass). We get $\dot{\varepsilon}_{\text{rot}} \approx 5 \cdot 10^{38}$ erg/s.

The proximity of the estimates of the magnetodipole emission of a pulsar and the decrease in the mechanical rotation energy testifies to the appropriateness of the model. The observed luminosity from the Crab-like nebula is $\approx 4 \cdot 10^{37}$ erg/s in the X-ray range and $\approx 2 \cdot 10^{36}$ erg/s in the optical one. These data do not contradict the model and indicate that about 10% of the energy of the long-wave primary emission is transformed into secondary X-ray emission in a plasma surrounding the star.

6.5. Calculate the electromagnetic potentials and the field strengths at the distances which satisfy the condition $l \ll r \ll \lambda$. Consider the electric dipole, quadrupole, and magnetic dipole terms.

Answer.

$$\varphi(\vec{r}, t) = \frac{q}{r} + \frac{\vec{p}(t') \cdot \vec{r}}{r^3} + \frac{Q_{\alpha\beta}(t')x_\alpha x_\beta}{2r^5};$$

$$\vec{E}(\vec{r}, t) = -\nabla\varphi(\vec{r}, t); \quad \vec{A}(\vec{r}, t) = \frac{\dot{\vec{p}}(t') \cdot \vec{r}}{cr} + \frac{\vec{m}(t') \times \vec{n}}{r^2};$$

$$\vec{H}(\vec{r}, t) = \nabla \times \vec{A} = \frac{\dot{\vec{p}} \times \vec{n}}{cr^2} + \frac{3\vec{n}(\vec{m} \cdot \vec{n})}{r^3} - \frac{\vec{m}}{r^3}.$$

The electric field is presented by static formulas with the time-dependent dipole and quadrupole moments. The formula for the magnetic field contains the additional term related to the Biot–Savart law: if the elementary dipole $\vec{p}(t) = e(t)\vec{l}$, then $\vec{H}_{\text{BS}} = \dot{\vec{p}} \times \vec{n}/cr^2 = J(t)[\vec{l} \times \vec{r}]/cr^3$ is the field of the elementary current $J(t) = \dot{q}(t)$, which flows in the segment l.

6.6. Find the equations for force lines of the electric and magnetic fields of an electric oscillator. Trace the qualitative changes in the field pattern in a zone adjacent to the oscillator and in the wave zone.

Answer. The magnetic force lines look like circles, whose planes are perpendicular to the axis z, and their centers lie on this axis. The electric force lines are described by the equations

$$C_1 = \sin^2 \vartheta \left[\frac{1}{r} \cos(kr - \omega t) + k \sin(kr - \omega t) \right], \quad C_2 = \alpha,$$

where C_1, C_2 are constants.

6.7. Find the formula for the momentum loss per unit time $-d\vec{L}/dt$ by a system that radiates as an electric dipole.

Answer. The momentum flow density

$$\Re = \frac{[\vec{n} \times \ddot{\vec{p}}](\vec{n} \cdot \dot{\vec{p}})}{2\pi c^3 r^2}.$$

To calculate the quantity $-\frac{d\vec{L}}{dt} = \int \Re r^2 d\Omega$, it is convenient to use the formula $\overline{n_i n_k} = \frac{1}{3}\delta_{ik}$.

As a result, we have:

$$-\frac{d\vec{L}(t)}{dt} = \frac{2}{3c^2} \dot{\vec{p}} \times \ddot{\vec{p}} \Big|_{t'=t-\frac{r}{c}}.$$

6.8. Study the influence of interference on the emission of electromagnetic waves by the following system of charges: two identical electric charges e move uniformly with nonrelativistic velocity and a frequency ω along a circular orbit with radius a, being on the opposite ends of a diameter. Find the polarization, angular distribution $\overline{dI/d\Omega}$, and emission intensity $\bar{I}$. How will the emission intensity be changed if one of the charges is removed?

Answer. $\vec{p} = \vec{m} = 0$, $\vec{Q} \neq 0$;

$$\vec{H} = \frac{1}{c} \ddot{\vec{A}} \times \vec{n} = -\frac{4ea^2\omega^3}{c^3 r} \sin \vartheta [\vec{e}_\vartheta \cos(2\omega t' - 2\alpha)$$
$$+ \vec{e}_\alpha \cos \vartheta \sin(2\omega t' - 2\alpha)].$$

The frequency of oscillations of the distributions of charge and current and, respectively, the frequency of the field exceed twice the rotation frequency ω of each of the charges along the orbit. The emission polarization is elliptic. It becomes a circular one if $\vartheta \to 0$ and π and passes to a linear one if $\vartheta = \frac{\pi}{2}$;

$$\frac{\overline{dI}}{d\Omega} = \frac{2e^2 a^4 \omega^6}{\pi c^5} \sin^2 \vartheta \, (1 + \cos^2 \vartheta); \quad \overline{I} = \frac{32}{5} \frac{e^2 a^4 \omega^6}{c^5}.$$

If one of the charges is removed, then the emission intensity increases by a factor of $(\frac{\lambda}{a})^2$, i.e., quite significantly, if $\frac{a}{\lambda} \ll 1$.

6.9. Under the condition $l \ll \lambda$, calculate the vector potential of the radiating system in the wave zone with regard for terms of the order (l/λ^2). Calculate the emission intensity, by assuming $\dddot{\vec{p}} \neq 0$, $\dddot{\vec{m}} \neq 0$, $\dddot{Q}_{\alpha\beta} \neq 0$.

Answer.

$$\vec{A}(\vec{r}, t) = \frac{\dot{\vec{p}}}{cr} + \frac{\dot{\vec{m}} \times \vec{n}}{cr} + \frac{\ddot{\vec{Q}}}{6c^2 r} + \frac{\vec{n}}{6c^2 r} \int r'^2 \dot{\rho} dV'$$

$$+ \frac{1}{2c^3 r} \int \vec{r}'(\vec{n} \cdot \vec{r}')^2 \dddot{\rho} dV'$$

$$- \frac{1}{c^3 r} \int \vec{r}'(\vec{n} \cdot \vec{r}')(\vec{n} \cdot \ddot{\vec{j}}) dV';$$

$$I = \frac{2}{3c^3}(\ddot{\vec{p}}^2 + \ddot{\vec{m}}^2) + \frac{1}{180c^5} \dddot{Q}_{\alpha\beta}^2 + \frac{2}{15c^5} \ddot{\vec{p}} \cdot \dddot{\vec{L}},$$

where

$$\vec{L} = \int \dot{\rho} r'^2 \vec{r}' dV' + \int [r'^2 \vec{j} - 3\vec{r}'(\vec{r}' \cdot \vec{j})] dV'.$$

The other notations are commonly accepted, and all time-dependent quantities are taken at the retarded time, $t' = t - r/c$.

6.10. Find the formulas for the strengths of the electromagnetic fields of electric $\vec{p}$ and magnetic $\vec{m}$ dipole oscillators in the vector form at large distances as compared with their sizes.

Hint. While differentiating with respect to $\vec{r}$, take into account that the dipole moments should be taken at the retarded time moment $t' = t - r/c$. Hence, they depend on r.

Answer. The field of a magnetic dipole:

$$\vec{E}_m(\vec{r}, t) = \frac{1}{c}\dot{\vec{A}}_m = \frac{\vec{n} \times \ddot{\vec{m}}(t')}{c^2 r} + \frac{\vec{n} \times \dot{\vec{m}}(t')}{c^2 r},$$

$$\vec{H}_m(\vec{r}, t) = \text{rot } \vec{A}_m$$

$$= \frac{3\vec{n}(\vec{m} \cdot \vec{n}) - \vec{m}}{r^3} + \frac{3\vec{n}(\dot{\vec{m}} \cdot \vec{n}) - \dot{\vec{m}}}{cr^2} + \frac{\vec{n} \times [\vec{n} \times \ddot{\vec{m}}]}{c^2 r}.$$

The field of an electric dipole can be obtained from the field of a magnetic dipole by means of the change $\vec{m} \to \vec{p}, \vec{H}_m \to \vec{E}_e, \vec{E}_m \to \vec{E}_e$.

6.11. A particle with charge q moves along a circle with radius a with velocity v (see Fig. 6.2).

(a) Find the scalar potential φ at the center of the circle;

(b) Determine the vector potential $\vec{A}$ at the center of the circle.

Answer. The electric and magnetic fields of a moving charge are described by the Liénard–Wiechert potentials

$$\varphi = \frac{q}{R' - (\vec{v}' \cdot \vec{R}')/c}, \qquad \vec{A} = \frac{qv'}{c[R' - (\vec{v}' \cdot \vec{R}')/c]},$$

where $\vec{R}'$ and $\vec{v}'$ are taken at $t' = t - R'/c$.

(a) During the motion along the circle, $\vec{v} \perp \vec{R}$ at any time. Therefore, we have $\varphi = \frac{q}{a}$ at the center of the circle.

(b) Analogously, $\vec{A} = \frac{q\vec{v}'}{ca}$, where $v'[t'] = v[t - (a/c)]$.

Let we start to count the time at the time moment, when the charge is located at a point P. We have

$$v'_x = -v\sin\left(\omega t - \frac{v}{c}\right); \qquad v'_y = v\cos\left(\omega t - \frac{v}{c}\right),$$

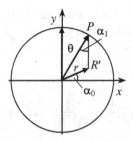

Fig. 6.2

$(\omega = v/a)$, i.e.,

$$A_x = -\frac{qv}{ca}\sin\left(\omega t - \frac{v}{c}\right); \quad A_y = \frac{qv}{ca}\cos\left(\omega t - \frac{v}{c}\right).$$

6.12. Deduce the formula for the potentials of a uniformly-moving charge, with the help of the relativistic transformation of a static Coulomb field (Liénard–Wiechert potentials):

$$\vec{A} = \frac{\vec{v}}{c\left(R - \frac{v\vec{R}}{c}\right)}; \quad \varphi = \frac{e}{R - \frac{v\vec{R}}{c}},$$

where $\vec{R} = \vec{r} - \vec{r}_0(t'); \vec{r}$ and $\vec{r}_0(t')$ are the radii-vectors of the observation point and the position of the charge at the moment of the emission

$$t' = t - \frac{|\vec{r} - \vec{r}_0(t')|}{c}.$$

6.13. Using the Liénard–Wiechert potentials (Problem 6.12), determine the strengths of the electric and magnetic fields of a moving charge.

Answer.

$$\vec{E} = e\frac{\left(\vec{R} - R\frac{\vec{v}}{c}\right)\left(1 - \frac{v^2}{c^2}\right)}{\left(R - \frac{\vec{R}\vec{v}}{c}\right)^3} + \frac{e}{c}\frac{\left[\vec{R} \times \left[\left(\vec{R} - \frac{R\vec{v}}{c}\right) \times \vec{v}\right]\right]}{\left(R - \frac{\vec{R}\vec{v}}{c}\right)^3},$$

$$\vec{B} = \frac{[\vec{R} \times \vec{E}]}{Rc}.$$

6.14. Find the law of transformation of components of the electric and magnetic fields in vacuum into the system which moves with velocity $\vec{v}$.

Answer. $\vec{E}'_{\parallel} = \vec{E}_{\parallel}, \vec{H}'_{\parallel} = \vec{H}_{\parallel}$;

$$\vec{E}'_{\perp} = \frac{\vec{E}_{\perp} + [\vec{v}, \vec{B}]}{(1 - v^2/c^2)^{1/2}}; \quad \vec{H}'_{\perp} = \frac{\vec{H}_{\perp} - [\vec{v}, \vec{B}]}{(1 - v^2/c^2)^{1/2}}.$$

Section 7

Emission of Electromagnetic Waves

7.1. Dipole and quadrupole emissions

The study of emission is simplified if the duration of the propagation of electromagnetic perturbations is much less that the characteristic time T related to the periodic motion of charged particles in the system,

$$l/c \ll T,$$

where T is the period;

$$l \ll \lambda,$$

where λ is the emission wavelength:

$$v \ll c.$$

We now expand formula (6.4) in a series in the size of the system. The current density takes the form

$$\vec{j}\left(\vec{r}', \, t - \frac{r}{c} + \frac{\vec{n} \cdot \vec{r}'}{c}\right)$$

$$= \vec{j}\left(\vec{r}', \, t - \frac{r}{c}\right) + \dot{\vec{j}}\left(\vec{r}', \, t - \frac{r}{c}\right)\frac{\vec{n} \cdot \vec{r}'}{c} + \dots \, .$$

The electric dipole emission is described by

$$\vec{A}(\vec{r}, \, t) = \frac{1}{cr} \int \vec{j}\left(\vec{r}', \, t - \frac{r}{c}\right) dV'.$$

Let us use the identity:

$$\vec{a} \cdot \vec{j} = \vec{j} \cdot \nabla'(\vec{a} \cdot \vec{r}') = \nabla' \cdot |\vec{j}(\vec{a} \cdot \vec{r}')| - \vec{a} \cdot \vec{r}'(\nabla' \cdot \vec{j}),$$

where $\vec{a}$ is a constant vector. We have

$$\vec{a} \int \vec{j} \left(\vec{r}', \, t - \frac{r}{c} \right) dV' = \vec{a} \frac{\partial}{\partial t} \int \vec{r}' \rho \left(\vec{r}', \, t - \frac{r}{c} \right) dV';$$

$$\vec{A}(\vec{r}, t) = \frac{\dot{\vec{p}}(t - r/c)}{cr}. \tag{7.1}$$

The emission intensity of a dipole takes the form

$$I = \frac{2\ddot{\vec{p}}^2}{3c^3},$$

where $\vec{p}$ is the dipole moment of the system.

The emission of one charged particle is described by the Larmor formula

$$I = \frac{2e^2 \dot{\vec{v}}^2}{3c^3},$$

where $\vec{v}$ is the velocity of a particle.

We note that a particle that moves with acceleration can radiate. The emission intensity for periodic motion is as follows:

$$I_m = \frac{4\omega_m^4}{3c^3} |\vec{p}_m|^2.$$

The quadrupole emission

$$\vec{p} = \sum_a e_a \vec{r}_a = \eta \sum_a m_a \vec{r}_a = \eta \vec{R} \sum_a m_a,$$

where $\vec{R}$ is the radius-vector of the center-of-mass.

The differential intensity of the quadrupole emission

$$\frac{dI}{d\Omega} = \frac{1}{4\pi c^3} \left[\ddot{\vec{m}} \times \vec{n} \right]^2 + \frac{1}{144\pi c^5} \left[\dddot{\vec{Q}} \times \vec{n} \right]^2 - \frac{1}{12\pi c^4} \ddot{\vec{m}} \cdot \left[\dddot{\vec{Q}} \times \vec{n} \right].$$

The total intensity over all directions equals

$$I = \frac{2}{3c^3}\dddot{m}^2 + \frac{1}{180c^5}\dddot{Q}_{\alpha\beta}^2,$$

where $\vec{Q}_{\alpha\beta}$ is the quadrupole moment, and $\vec{Q}$ is the convolution ($\vec{Q}_\alpha = \vec{Q}_{\alpha\beta} \cdot n_\beta$).

Hertz vector and the emission of an antenna. The Hertz vector is connected with the electromagnetic potentials by the relations:

$$\varphi = -\operatorname{div} \vec{Z}^{(e)}; \quad \vec{A} = \frac{1}{c}\frac{\partial \vec{Z}^{(e)}}{\partial t}.$$

The distribution of charges and currents:

$$\rho = -\operatorname{div} \vec{P}; \quad \vec{j} = \frac{1}{c}\frac{\partial \vec{P}}{\partial t}. \tag{7.2}$$

The total charge is zero:

$$q = \int_{V\to\infty} \rho\, dV = -\oint_{S\to\infty} \vec{P}\cdot d\vec{S} = 0.$$

The Hertz vector is described by the d'Alembert equation

$$\Delta\vec{Z}^{(e)} - \frac{1}{c^2}\frac{\partial^2 \vec{Z}^{(e)}}{\partial t^2} = -4\pi\vec{P}. \tag{7.3}$$

The solution of this level is as follows:

$$\vec{Z}^{(e)}(\vec{r},t) = \int \frac{\vec{P}(\vec{r}',t - |\vec{r}-\vec{r}'|/c)}{|\vec{r}-\vec{r}'|}\, dV'.$$

$$* * *$$

Example 7.1. A particle with charge e oscillates along the axis Oz by the law $z(t) = a \sin \omega t$. Calculate the emission intensities $dI_m/d\Omega$ at multiple frequencies $\omega_m = m\omega$, $m = 1, 2, \ldots$, not assuming that the ratio a/λ is small.

Solution. We use the expansion of the potentials in Fourier series and calculate the Fourier harmonic of the magnetic field strength. We obtain

$$\vec{H}_m = i\frac{em\omega^2}{2\pi c^2}\frac{e^{i\vec{k}_m r}}{r}\int_0^T \vec{n}\times\vec{v}(\tau)\exp i(m\omega\tau - \vec{k}_m\cdot\vec{s}(\tau))d\tau,$$

where $\vec{k}_m = m\omega\vec{n}/c, \vec{s}(\tau) = \vec{e}_z z(\tau)$. The integral over the time is expressed in terms of the Bessel function.

We have

$$\frac{dI_m}{d\Omega} = \frac{c}{2\pi}|\vec{H}_m|^2 r^2 = \frac{e^2\omega^2}{2\pi c}\tan^2\theta m^2 J_m^2(m\beta\cos\theta),$$

where $\beta = a\omega/c, \theta$ is the angle between the direction of emission n and the axis Oz.

Here, we took into account that the harmonics with numbers m and $-m$ make the same contribution to the emission.

If $\beta \ll 1$, we obtain

$$\frac{dI_m}{d\Omega} \approx \frac{e^2 a^2\omega^4}{8\pi c^3}\sin^2\theta$$

(dipole emission);

$$\frac{dI_2}{d\Omega} \approx \frac{e^2 a^4\omega^6}{2\pi c^5}\sin^2\theta\cos^2\theta$$

(the quadrupole emission), which contains the small factor $(a\omega/c)^2$, as compared with the dipole one.

Example 7.2. In a linear antenna l in length, the standing wave I with amplitude I_0, frequency ω, and nodes at the antenna ends is excited. The number of half-waves of the current, which are placed on the antenna length, is m. Find the angular distribution of the emission $\frac{\overline{dI}}{d\Omega}$.

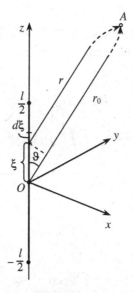

Fig. 7.1

Solution. We choose the coordinate system shown in Fig. 7.1. The distribution of the current over the antenna is given by the formula

$$I = I_0 \sin\left[k\left(\xi + \frac{l}{2}\right)\right] e^{-i\omega t},$$

where $k = \frac{\omega}{c} = \frac{m\pi}{l}$.

The electric dipole moment of a unit length of the antenna $P = \frac{i}{\omega}I$, according to Eq. (7.2). An element $d\xi$ of the antenna can be considered as an electric dipole oscillator with moment $dp = P\, d\xi$. Since the inequality $d\xi \ll \lambda$ is satisfied, the magnetic field created by the element $d\xi$ at a point A can be calculated by formula (7.1):

$$d\vec{H}_0(\vec{r}_0,\, t) = \frac{\omega^2}{c^2 r}\vec{e}_\alpha \sin\, \vartheta P\left(t - \frac{r}{c}\right) d\xi,$$

where

$$r = r_0 - \xi \cos\, \vartheta.$$

Since we need to determine only the field in the wave zone, the quantity $\frac{\sin\vartheta}{r}$, which varies slightly in the domain $r \gg l$, can be transferred outside the integral.

Thus,

$$H_r = H_\vartheta = 0;$$

$$H_\alpha = -\frac{i\omega\sin\vartheta}{c^2 r_0}I_0\, e^{i(kr_0-\omega t)}\int_{-\frac{1}{2}}^{\frac{1}{2}} e^{ik\xi\cos\vartheta}\sin\, m\pi\left(\frac{\xi}{l}-\frac{1}{2}\right)d\xi.$$

Executing the integration, we find the angular distribution by the formula $\frac{dI}{d\Omega} = \frac{c}{4\pi}\overline{H_\alpha^2}r_0^2$:

$$\frac{dI}{d\Omega} = \begin{cases} \dfrac{I_0^2}{2\pi c}\cdot\dfrac{\cos^2\left(\frac{m\pi}{2}\cos\vartheta\right)}{\sin^2\vartheta}, & \text{if } m \text{ is odd,} \\[4mm] \dfrac{I_0^2}{2\pi c}\cdot\dfrac{\sin^2\left(\frac{m\pi}{2}\cos\vartheta\right)}{\sin^2\vartheta}, & \text{if } m \text{ is even.} \end{cases}$$

The character of the angular distribution is seen from the polar diagrams in Fig. 7.2. The dashed line shows the current distribution over the antenna length, and the solid line gives the angular distribution of the emission.

Example 7.3. Let a standing wave of current of the form $I = I_0\sin n\alpha' e^{-i\omega t}$ be excited in a circular wire loop l in length with radius a. Determine the electromagnetic field $\vec{H}, \vec{E}$ in the wave zone.

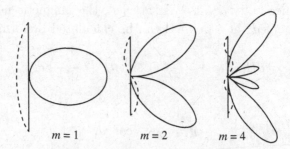

$m = 1$ $m = 2$ $m = 4$

Fig. 7.2

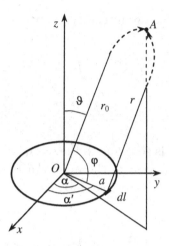

Fig. 7.3

Solution. If the distance r of the observation point $A(r_0, \vartheta, \alpha)$ (Fig. 7.3) from the loop is large ($r \gg a$), then we may consider that the radii-vectors $\vec{r}$ of all elements $d\vec{l}$ of the loop are parallel, and

$$r_0 - a \cos \varphi = r_0 - a \sin \vartheta \cos(\alpha' - \alpha).$$

The element $d\vec{l}$ has the electric dipole moment $d\vec{p} = P d\vec{l} = \frac{i}{\omega} I d\vec{l}$, where P is the electric dipole moment of a unit length of the wire and creates the magnetic field:

$$d\vec{H}(\vec{r}_0, t) = \frac{\omega^2}{c^2} \frac{d\vec{p}(t')\vec{n}}{r} = -i\frac{\omega a}{c^2} \frac{I_0}{r_0} e^{-i\omega t + ikr_0 - iak \sin \vartheta \cos(\alpha' - \alpha)}$$
$$\times \sin n\alpha'[\cos(\alpha' - \alpha)\vec{e}_\vartheta + \cos \vartheta \sin(\alpha' - \alpha)\vec{e}_\alpha]d\alpha' \quad (7.4)$$

at the point A. In the denominator of expression (7.4), we neglect the quantity of the order of a, as compared with r_0.

The problem of determination of the field is reduced to the integration:

$$H_\vartheta = \frac{i\omega a}{c^2} \frac{I_0}{r_0} e^{i(kr_0 - \omega t)} \int_{-\pi}^{\pi} \cos(\alpha' - \alpha) \sin n\alpha' e^{0 - ika \sin \vartheta \cos(\alpha' - \alpha)} d\alpha'.$$

Let us introduce of the integration variable $\beta = \alpha' - \alpha$. We get

$$H_\vartheta = -\frac{i\omega a}{c^2}\frac{I_0}{r_0}e^{i(kr_0-\omega t)}\left(\cos n\alpha \int_{-\pi}^{\pi}\cos\beta\sin n\beta e^{0-ika\sin\vartheta\cos\beta}d\beta\right.$$

$$\left. + \sin n\alpha \int_{-\pi}^{\pi}\cos\beta\cos n\beta e^{0-ika\sin\vartheta\cos\beta}d\beta\right).$$

The first integral is zero due to the oddness of the integrand. The second can be transformed to the interval $(0, \pi)$ (integrand is even) and be presented in terms of the derivative of a Bessel function. Thus,

$$H_\vartheta(r_0, t) = -E_\alpha = \frac{2\pi\omega a}{c^2}\frac{I_0}{r_0}e^{i\left(kr_0-\omega t-n\frac{\pi}{2}\right)}\sin n\alpha J_n'(ka\sin\vartheta).$$

Performing the analogous calculations with the use of the formula $J_{n-1}(x) + J_{n+1}(x) = \frac{2n}{x}J_n(x)$, we get

$$H_\alpha(r_0,\ t) = E_\vartheta = \frac{2\pi\omega a I_0 e^{i\left(kr_0-\omega t-n\frac{\pi}{2}\right)}}{c^2 r_0}\cos n\alpha\frac{J_n(ka\sin\vartheta)}{ka\tan\vartheta}.$$

Example 7.4. Let the decreasing current

$$J = J_0\sin[k_m(\xi + l/2)]e^{-\gamma t}\cos\omega_m t\Theta(t), \quad -l/2 \le \xi \le k/2,$$

flow in a linear antenna (see Example 7.2). Here, $\omega_m = ck_m = m\pi c/l$, Θ is a step-function; $m = 1, 2, \ldots$ is the number of half-waves, which are placed on the antenna length.

Calculate the spectral density of the emission $\frac{d^2 I_\omega}{d\omega d\Omega}$ in the symmetry plane perpendicular to the antenna axis.

Solution. We introduce the electric Hertz vector according to Eq. (7.3) and calculate the magnetic field:

$$\vec{H} = \operatorname{rot}\frac{1}{c}\frac{\partial\vec{Z}^{(e)}}{\partial t} = \operatorname{rot}\frac{1}{c}\int\frac{\dot{\vec{P}}(\vec{r}',t - R/c)}{R}dV'$$

$$= \text{rot}\frac{\vec{e}_z J_0}{c} \int_{-l/2}^{l/2} \frac{1}{R} \sin[k_m(\xi + l/2)] \cos \omega_m(t - R/c)$$

$$\times \exp[-\gamma(t - r/c)]d\xi.$$

Using Fig. 7.1, we get $R \approx r_0 - \xi \cos \vartheta \approx r_0$, since $\vartheta = \pi/2$ in the symmetry plane. By calculating the integral with respect to $d\xi$, we will obtain a nonzero result only for odd m. After the calculation of a curl and the Fourier integral for odd m, we get the Fourier component of the magnetic field

$$\vec{H}_\omega = \frac{J_0[\omega_m^2 + \gamma(i\omega - \gamma)]}{r_0 c^2 k_m[\omega_m^2 + (i\omega - \gamma)^2]} = [\vec{n}_0 \times \vec{e}_z]e^{i\omega r_0/c}$$

and the spectral density of the emission in the symmetry plane $\vartheta = \pi/2$:

$$\frac{d^2 I_\omega}{d\omega d\Omega} = \frac{J_0^2[(\omega_m^2 - \gamma^2)^2 + \gamma^2\omega^2]}{4\pi^2 c\omega_m^2[(\omega_m^2 - \omega^2 + \gamma^2)^2 + 4\gamma^2\omega^2]}.$$

For m even, the spectral density of the emission in the symmetry plane is zero. If $\gamma^2 \ll \omega_m^2$, the emission spectrum has a typical resonance shape with a sharp maximum at the frequency $\omega^2 = \omega_m^2 + \gamma^2$.

Example 7.5. Find the formulas for the electric dipole $\vec{Z}_p$, quadrupole $\vec{Z}_Q$, and magnetic dipole $\vec{Z}_m$ terms of the expansion of the electric Hertz vector, which should be valid for any time dependence of currents and charges. These formulas should be proper at the distances $r \gg a$, $\lambda \gg a$ (the validity of the condition $r \gg \lambda$ is not obligatory). Here, a is the size of the system. The upper index (e) of the Hertz vector is omitted.

Solution. We expand the Hertz vector $\vec{Z}(\vec{r}, t)$ into monochromatic components. We get

$$\vec{Z}_p(\vec{r}, t) = \frac{\vec{p}(t')}{r},$$

where $t' = t - \frac{r}{c}$;

$$\vec{Z}_Q(\vec{r}, t) = \frac{1}{2r^2}\dot{Q}(t') + \frac{1}{2rc}\ddot{Q}(t');$$

$$\vec{Z}_m(\vec{r}, t) = \frac{\dot{\vec{m}}(t') \times \vec{n}}{r} + \frac{c}{r^2}\left[\int \vec{m}(t')\,dt'\right]\vec{n}. \qquad (7.5)$$

These formulas are valid for $r \gg a$, where a is the size of the system. The arbitrary constant, which arises at the calculation of the integral in (7.5), does not affect the strengths of the field.

Problems

7.1. An electric and magnetic dipole are mutually perpendicular, oscillate with frequency ω_0, and are located at the same point of the space. Find the time-averaged angular distribution $\frac{\overline{dI}}{d\Omega}$ and total emission intensity $\bar{I}$.

Answer.

$$\frac{dI}{d\Omega} = \frac{\omega_0^4}{4\pi c^3}\{p^2(1 - \sin^2\vartheta\cos^2\alpha)$$

$$+ m^2\sin^2\vartheta + mp\sin\vartheta\sin\alpha\};$$

$$I = \frac{2\omega_0^4}{3c^3}(p^2 + m^2).$$

Here, we used the coordinate system whose axis x is directed along $\vec{p}$, and the axis z is parallel to $\vec{m}$.

By averaging the emission intensity over the period of oscillations, we obtain

$$\frac{\overline{dI}}{d\Omega} = \frac{\omega_0^4}{8\pi c^3}\{p_0^2(1 - \sin^2\vartheta\cos^2\alpha) + m_0^2\sin^2\vartheta\};$$

$$\bar{I} = \frac{\omega_0^4}{3c^3}(p_0^2 + m_0^2).$$

7.2. A uniformly charged (over volume) drop pulsates so that its density is constant. In this case, the drop surface is described by the equation

$$R(\vartheta) = R_0[1 + aP_2(\cos\vartheta)\cos\omega t],$$

where $a \ll 1$.

Let the drop charge be q. Find the time-averaged angular distribution $\frac{\overline{dI}}{d\Omega}$ and total intensity emission $\overline{I}$.

Answer.

$$\frac{\overline{dI}}{d\Omega} = \frac{9}{800\pi}\frac{\omega^6 q^2 R_0^4 a^2}{c^5}\sin^2\vartheta\cos^2\vartheta;$$

$$\overline{I} = \frac{3}{500}\frac{\omega^6 q^2 R_0^4 a^2}{c^5}.$$

7.3. Find the total emission intensity $\overline{I}$ and the resistance $R = \frac{2\overline{I}}{I_0^2}$ of the antenna from Problem 7.2.

Hint. The result is presented in terms of the integral cosine

$$\mathrm{Ci}(x) = C + \ln x + \int_0^\pi \frac{\cos t - 1}{t}dt,$$

where $C = 0.577\ldots$ is the Euler constant.

Answer.

$$\overline{I} = \frac{I_0^2}{2c}[\ln(2\pi m) + C - \mathrm{Ci}(2\pi m)];$$

$$R = 2\frac{\overline{I}}{I_0^2} = \frac{1}{c}[\ln(2\pi m) + C - \mathrm{Ci}(2\pi m)].$$

7.4. Let the running wave of the current $I = I_0 e^{i(k\xi - \omega t)}$ propagate in a linear antenna l in length. Here, $k = \frac{\omega}{c}$, and ξ is the coordinate of a point on the antenna. Find the angular distribution $\frac{\overline{dI}}{d\Omega}$ and the total emission intensity $\overline{I}$.

Answer.

$$\frac{\overline{dI}}{d\Omega} = \frac{I_0^2}{2\pi c}\frac{\sin^2\vartheta\sin^2\left[\frac{kl}{2}(1 - \cos\vartheta)\right]}{(1 - \cos\vartheta)^2};$$

$$\overline{I} = \frac{I_0^2}{c}\left[C - 1 + \ln\frac{4\pi l}{\lambda} - \mathrm{Ci}\left(\frac{4\pi l}{\lambda}\right) + \frac{\sin\left(\frac{4\pi l}{\lambda}\right)}{\frac{4\pi l}{\lambda}}\right],$$

where $\lambda = \frac{2\pi}{k}$ is the emitted wave length, and ϑ is a polar angle counted from the coordinate axis ξ.

It is easy to verify that the running wave emits with higher intensity than a standing wave with the same values of $l, \lambda, \mathfrak{J}_0$.

7.5. Let N antennas parallel to the axis Ox be placed in the plane xz. Each antenna has length l.

The distance between the adjacent antennas a is given. In each antenna, the current $J = J_0 \sin k_m z \cos \omega_m t$ flows. Find the angular distribution of the time-averaged emission intensity $\frac{\overline{dI}}{d\Omega}$.

Answer. The magnetic field is created by N sources. Calculating it in terms of the Hertz vector, we obtain

$$\vec{H}(\vec{r}_0, t) = -\text{Re}\left\{ i\frac{\vec{n}_0 \times \vec{e}_z J_0 k_m}{c r_0} \exp[i\omega_m(t - R/c)] \right.$$

$$\times \sum_{s=0}^{N-1} \exp[isk_m a \sin\vartheta \cos\varphi]$$

$$\left. \times \int_{-l/2}^{l/2} \sin[k_m(\xi + l/2)] \exp[ik_m \xi \cos\vartheta]d\xi \right\}.$$

The time-averaged emission intensity is given by the formula

$$\frac{\overline{dI}}{d\Omega} = \frac{c r_0^2}{8\pi}|\vec{H}|^2 = \frac{J_0^2}{2\pi c \sin^2\vartheta}$$

$$\times \frac{\sin^2\left[\frac{N}{2}k_m a \sin\vartheta \cos\varphi\right]}{\sin^2\left[\frac{1}{2}k_m a \sin\vartheta \cos\varphi\right]} \left\{ \begin{array}{c} \cos^2\left[\frac{m\pi}{2}\cos\vartheta\right] \\ \sin^2\left[\frac{m\pi}{2}\cos\vartheta\right] \end{array} \right\}$$

where the upper and lower quantities in the braces correspond to odd and even m, respectively.

7.6. Calculate the flow of Poynting's vector $\vec{\gamma}$ through the surface of a long straight wire with the resistance R per unit length. To compare the obtained result with ohmic losses.

Answer. The energy flow density, which is equal to the Poynting's vector modulus, is

$$\vec{\gamma} = \frac{c}{4\pi}\vec{E} \times \vec{B} = (c^2/R)\vec{I} \times \vec{B}.$$

The energy flow is directed perpendicularly to the wire surface. Integrating over the surface of a wire with unit length, we get

$$\int \vec{\gamma} d\vec{s} = I^2 R.$$

Hence, the flow of Poynting's vector through the surface of a straight wire is equal to the ohmic losses in this wire.

7.7. A long coaxial cable is fabricated of two concentric cylinders, which conduct a current. One end of the cable is connected with an electric battery with voltage V. Another end of the cable is connected with a resistance R. Using Poynting's vector, calculate the energy flow velocity.

Answer.

$$|\vec{\gamma}| = \frac{c}{4\pi}|\vec{E} \times \vec{B}| = \frac{IV}{2\pi r^2 \ln(b/a)};$$

$$w = \frac{1}{8\pi}(E^2 + B^2) = V \frac{2}{16\pi r^2}\left[\frac{1}{\ln^2(b/a)} + \frac{1}{R^2 c^4}\right].$$

7.8. A plane electromagnetic wave falls onto a free electron and forces it to oscillate. Find the ratio of the energy emitted by the electron per unit time to the energy flow density of the incident electromagnetic wave. The wave frequency is assumed small. Therefore, the influence of the magnetic field $\vec{B}$ of the wave on the motion of the electron can be neglected.

Answer. By dividing the flow into the energy density of the incident wave $J = \varepsilon_0 c E^2$, we obtain the differential cross-section of scattering of the electromagnetic wave by the electron

$$d\sigma = \frac{dJ}{J} = \left(\frac{q^2}{mc^2}\right)^2 \sin^2 \theta d\Omega.$$

The total scattering cross-section

$$\sigma = \frac{8\pi}{3} \left(\frac{e^2}{mc^2} \right)^2.$$

7.9. Let N turns of a wire be wound on a toroid, whose mean radius is R, and the transverse cross-section radius is $r (r \ll R)$. At the moment in time $t = 0$, a current starts to flow along the wire, whose intensity increases with time by a linear law

$$I(t) = Kt.$$

(a) Find the magnetic field and the energy stored in the toroid at time t.

(b) Find the direction and the value of Poynting's vector at any internal point of the toroid at time t.

Answer.

(a) $B = -\dfrac{2NKt}{c^2 R}; \quad E = \dfrac{\rho NK}{c^2 R},$

$$w = \frac{1}{8\pi} \int (E^2 + B^2) dV$$

$$= \frac{\pi r^2}{c^2} \frac{N^2 K^2 t^2}{R} + \frac{\varepsilon_0}{2} \left(\frac{NK}{c^2 R} \right)^2 \pi r^5.$$

(b) The directions $\vec{E}$ and $\vec{B}$ are such that Poynting's vector $\vec{\gamma}$ at every point on the toroid surface is normal to the surface and is directed inwards. In addition, $\vec{E} \perp \vec{B}$ at every point. Hence, the modulus of Poynting's vector is

$$\vec{\gamma} = \frac{c}{4\pi} \vec{E} \times \vec{B} = \frac{2rt}{c^4 R^2} (NK)^2.$$

The total absorbed energy power

$$\oint \vec{\gamma} d\sigma = \frac{2\pi r^2 t (NK)^2}{c^4 R}.$$

7.10. Show that there is no dipole emission during the collision of two identical particles.

Answer. In the center-of-mass system, the dipole moment of the system

$$\vec{p} = e_1\vec{r}_1 + e_2\vec{r}_2 = \mu\left(\frac{e_1}{m_1} - \frac{e_2}{m_2}\right)r,$$

where μ is the reduced mass; r is the relative coordinate of the particles; m_1 and m_2 are the masses of the particles.

For identical particles, $e_1 = e_2$, $m_1 = m_2$. Then $\vec{p} = 0$, and, hence, no dipole emission, which is proportional to $|\dddot{\vec{p}}|^2$, happens.

7.11. Find the emission intensity of a particle with mass m, which moves along a circular orbit with radius a under the action of Coulomb forces. Express the answer in terms of the energy of the particle.

Answer.

$$J = \frac{e^6}{2c^3a^4m^2} = \frac{|\varepsilon|^4}{3m^2c^3e^2},$$

where ε is the energy of a particle.

7.12. Find the effective scattering cross-section for an elliptically polarized wave with frequency ω by an oscillator with mass m, charge e, eigenfrequency ω_0, and damping coefficient γ.

Answer. For the elliptically polarized wave, we write

$$\vec{E} = \vec{A}\cos\omega t + \vec{B}\sin\omega t,$$

where $\vec{A} \perp \vec{B}$.

Then the differential effective scattering cross-section is equal to

$$d\sigma = \left(\frac{e^2}{mc^2}\right)^2 \frac{\omega^1}{(\omega_0^2 - \omega^2)^2 + \omega^2\gamma^2} \frac{[\vec{A}\vec{n}]^2 + [\vec{B}\vec{n}]^2}{A^2 + B^2} d\Omega,$$

where $\vec{n}$ is a unit vector along the scattering direction.

7.13. Let a dipole with moment $\vec{p}$, which is located at the coordinate origin, oscillate with frequency ω. Let a particle, whose polarizability is β, be present at a point with radius-vector $\vec{d}(\vec{d} \perp \vec{p})$. Find the intensity of electromagnetic waves emitted by such system, by assuming that $d \ll \lambda$, where λ is the emission wavelength.

Answer.

$$j = \frac{p^2 \omega^4}{c^3} \left(1 - \frac{\beta}{4\pi d^3} \right)^2 .$$

7.14. Let a charged particle in the medium be decelerated by a force proportional to the velocity of motion of the particle. Find the energy emitted by the particle per unit time per unit solid angle. The initial velocity of motion is $\vec{v}_0$.

Answer.

$$\frac{dJ(\omega)}{d\Omega} = \frac{e^2 \vec{v}_0^2}{c^3} \frac{k^2}{k^2 + \omega^2} \sin^2 \vartheta .$$

References

1. L. D. Landau and E. M. Lifshitz, *The Classical Theory of Fields* (Pergamon Press, Oxford, 1983).
2. L. D. Landau, E. M. Lifshitz, *Electrodynamics of Continuous Media* (Pergamon Press, New York, 1984).
3. B. G. Levich, *Theoretical Physics: An Advanced Text*, Vol. 1 (North Holland, Amsterdam, 1970).
4. A. Sommerfeld, *Electrodynamics* (Acad. Press, New York, 1952).
5. I. E. Tamm, *Fundamentals of the Theory of Electricity* (Mir, Moscow, 1979).
6. A. Einstein, *The Meaning of Relativity* (Princeton Univ. Press, Princeton, 1956).
7. V. V. Batygin and I. N. Toptygin, *Collection of Problems of Electrodynamics* [in Russian] (Nauka, Moscow, 1970).
8. L. G. Grechko, V. I. Sugakov, O. F. Tomasevich, and A. M. Fedorchenko, *Collection of Problems of Theoretical Physics* [in Russian] (Vyssh. Shkola, Kiev, 1984).
9. I. S. Gradshtein and I. M. Ryzhik, *Tables of Integrals, Series, and Products* (Acad. Press, New York, 1980).
10. A. P. Prudnikov, Yu. A. Brychkov, and O. I. Marichev, *Integrals and Series* (Gordon and Breach, New York, 1986–1992).
11. R. Feynman, R. Leighton, M. Sands. *The Feynman Lectures on Physics. Exercises* (Addison-Wesley Publishing Com., London,1964).
12. S.Weinberg. *The Quantum Theory of Fields* (Cambridge Univ. Press, New York, 1995).

Printed in the United States
By Bookmasters